ASSEMBLAGES OF VIOLENCE IN EDUCATION

Assemblages of Violence: Everyday Trajectories of Oppression brings together fields including new materialisms, anthropology, curriculum theory, and educational foundations to examine how violence is intertwined with everyday events and ideas. Artfully weaving participant narratives in two contexts that exist a literal world apart—queer middle school youth of color in an urban context and Indian women who have survived domestic violence—Assemblages of Violence conceptualizes how social justice functions in opposition to normalized aggressions. Often overlooked, these deeply significant connections document how multiplicities of aggression operate as business-as-usual in a variety of spaces and places, including those that are often thought of as helpful. To these ends, this book introduces pathologies to theoretically and methodologically trace affects in order to more clearly perceive both where and how violence is embedded in and between sociopolitical and cultural ways of being, knowing, and doing. In so doing, Assemblages of Violence argues that pathologizing trajectories of violence can provide theoretical and methodological tools for those seeking to engage in a pedagogy of equity, access, and care to help people and communities in ways they wish to be helped.

Boni Wozolek is currently an Assistant Professor of Education at Penn State Abington. Her work considers questions of social justice, qualitative research methods, and teaching practices that focus on the examination of race, genders, and sexual orientations in schools.

ASSEMBLAGES OF VIOLENCE IN EDUCATION

Everyday Trajectories of Oppression

Boni Wozolek

Routledge
Taylor & Francis Group
NEW YORK AND LONDON

First published 2021
by Routledge
52 Vanderbilt Avenue, New York, NY 10017

and by Routledge
2 Park Square, Milton Park, Abingdon, Oxon, OX14 4RN

Routledge is an imprint of the Taylor & Francis Group, an informa business

Library of Congress Cataloging-in-Publication Data
A catalog record for this title has been requested

ISBN: 978-0-367-68897-4 (hbk)
ISBN: 978-0-367-40980-7 (pbk)
ISBN: 978-1-003-13951-5 (ebk)

Typeset in Times NR MT Pro
by KnowledgeWorks Global Ltd.

For Evalyn and Isaac—Your lives give me purpose.

For Savio and Beth—Your memory centers me in my past-future self.

TABLE OF CONTENTS

ACKNOWLEDGEMENTS

Se wo were fi na wosan kofa a yenki: It is not taboo to go back and fetch what you forgot.

The phrase above comes from the Twi language of Ghana. In its abbreviated form, the term "sankofa" means that a person must go back to her roots in order to move forward. Although acknowledgements usually come at the beginning, the act of recognizing the help, love, and support someone gave along the journey is truly an act of remembering, of moving back before stepping forward. With that in mind, I would like to thank the following people.

First, and most significantly, to my participants: This work would not exist without you. You bravely shared your stories in a way that rendered us both vulnerable. While sharing might not have always been easy, it was always appreciated.

Second, to my family: Adam, in the middle of a pandemic you helped me to carve out time so this project could come to fruition while listening patiently while I struggled with theory over dinner. For that, thank you. To my beautiful children, Evalyn and Isaac, you gave up (although not always willingly) bedtime stories and playtime with mom so I could make this happen. I owe so much of the woman and scholar I am to the opportunity of being your mom. For your patience, your unconditional love, and your support, I am forever grateful. To my parents, Amy, Baji, Steve, Dia, and Gary, my godmother Annie, and my siblings, Ryan, Jenna, Dina, and Vince, who kindly asked about book updates, thank you for your love and interest in my work.

Third, to my chosen family: Walter Gershon, without hesitation you listened to ideas and spent time thinking through drafts with me. This project would not be what it is without your kindness and support. Rob Helfenbein, your support encouraged me to start the writing process. Without "words on page" in my head, there are many times I might not have had the courage to begin or continue. To Micha Rahder, my gratitude for your professional editorial eye. To Vicente Ortiz Cortez, my appreciation for your beautiful art that has graced this cover. To Reagan Mitchell, your gentle care is always encouraging to my ways of being and keeps me going. To Rachael Eriksen Brown, Kathy Fadigan, and Ann Martinelli, thank you for your patience in listening to book updates that you never asked to hear. My gratitude to my administrative team at Penn State Abington, Fran Sessa, Andy August, and Damian Fernandez, for your continued support.

Finally, to the rest of mi familia: Sherick Hughes, Thea Berry, Fran Huckaby, Ming Fang He, Michael Boucher, Karen Burgard, Jamie Buffington-Adams, Amoaba Gooden, Jim Henderson, Tommy Ender, Aparna Tarc, Kent Den Heyer, Anneli Frelin, Denise Taliaferro Baszile, Sandro Barros, Roland Sintos Coloma, Ramon Goings, Kim Powell, Adell Cothorne, Raúl Mora, Cynthia Dillard, Ed Brockenbrough, Roland Mitchell, Erica Meiners, Gabe Huddleston, Sam Rocha, Cris Mayo, Nicholas Ng-A-Fook, Jerry Rosiek, David Lee Carlson, Greg and Michelle Dimitriadis, Silvia Bettez, Valerie Malpass, Stacy Williams, Johnny Lupinacci, Camika Royal, Abraham DeLeon, Lance McCready, Dave Stovall, Janet Miller, Michelle Herren, Rick Fish, Jean Clandinin, Morna McDermott, George Noblit, Deepak Subramony, and countless others who I'm sure I'll be kicking myself for forgetting at this moment—my career is on a particular trajectory because of your guidance and care. For that, and so much more, thank you.

PREFACE

> *"Birth stories are everywhere and nowhere. Seen in every movie theatre but heard only in brief gasps of attention in grocery store lines or parking lots, inculcated in prenatal classrooms but shamed to the edges of conversation, birth stories permeate and haunt our everyday lives."*
>
> —*Pollock, 1999*

Although discussing the birth of babies, Pollock's quote resonates strongly with the metaphorical birth of ideas. The birth of a book, for example, is often celebrated among colleagues and friends. We buy each other drinks, put each other up for awards, and converse about our new arrivals on social media and at conferences. What we rarely discuss is how they were conceived. Perhaps this is because the conceptualization of the book is often a much longer process, one that frequently takes decades to come to fruition before the proposal is submitted. In the life of a scholar, writing a book is like pulling one prominent note from the concert that is a career and attending to it before allowing it to fold back into the symphony of scholarly ideas.

This process makes it difficult to decipher the moment of arrival, when the ideas coalesced to begin gestation. Although authors often identify with the complex process of ideas being sewn together and pulled apart, this birthing process, like most processes of bringing something to life, is often thought of as a duality—in progress versus finished. However, I would like to attend to these moments of conception that are often unheard in our celebrations and congratulatory remarks. These moments are the lifeblood of our books, yet they exist, as Pollock (1999) reminds us, everywhere and nowhere. They are the scribbles on a bar napkin, the thoughts that wake us in the night, and that nagging idea that bothers your brain like a musical earworm—existing unnamed and yet trying desperately to spill out onto the page.

These moments are often ignored—or sometimes hidden from others and even from ourselves—until the book has arrived. Then we take a deep breath of relief, declaring our work done...until we turn to the next book, article, or presentation looming on the horizon. The problem with this duality is that everything is intertwined (Gottlieb, 2000). Much in the way one cannot remember the entirety of their wedding day, the day their child was born or came home, or the day they lost a love, the story is as much about a past colliding with the present and future as it is about the arrival. In short, ideas are still becoming, even after they have become. The process of arriving is therefore just as, if not more, significant than the moment of the arrival itself.

This book shares the entangled becoming of my colleagues' work, in that there were many moments where the twinkle of thoughts served as an already-there-yet-to-be (Pinar, 1998) set of possibilities that have become this text. In particular, I would like to focus on one of these instances. My daughter, Evalyn, loved to read as a toddler. My partner and I spent countless hours delightfully hunched on the floor next to her favorite stuffed bear, flipping through pages of children's books. Some of my most beautiful memories are of her engaged in the wonder that comes from sharing a familiar story with someone you love.

When she reached kindergarten, something shifted. She began to hate reading. By first grade, she was considered "below grade level". The teacher decided that the best way to fix this pedagogically constructed delay was to assign extra reading and writing packets. This was in addition to the weekly homework that was assigned to the rest of the class. Although these assignments could be understood as a form of help, Evalyn perceived them as a punishment. As a result of far too much homework employed in this success-through-punishment tactic, her daily assignments were often met with tears and anxiety. The days of my baby begging to hear a good story had evaporated, leaving only a trace of sorrow in my bones as I watched my child struggle with school.

One day my partner introduced Evalyn to audio books to help fill the long drive across state lines as we traveled to see family. It didn't take long before she was hooked. Although it began as a special treat in the car, it quickly turned into her begging to listen to stories before bed. Then in the morning as she was getting ready for school. Then at the dinner table. Soon we had to put limits on the length of time and when she could indulge in the experience.

After a while, audio books turned back into a love of printed materials. By the end of second grade, while she refused to finish the assessment readers at school that she complained "lacked a real story"—meaning she remained below grade level in the traditional sense of schooling—she was reading everything from *The Critter Club* to *What Makes a Baby*. She eagerly listened to the *Percy Jackson* series while checking out *The Red Pyramid* from her school library. As a mother who, like most parents, wanted to watch her child be happy, I felt like, for a moment, I could breathe.

One of our mother-daughter activities when we have had collective time to relax has been "girls' night". There are two rules that she and I decided on for girls' night: you need to get completely ready for bed before the festivities can start and you must, to the best of your ability, have fun. In celebration of my first tenure-track position, her success in school, and the fact that her baby brother was interested in sleeping early one evening, we decided a girls' night was just what this doctor had ordered.

While deciding on a movie to compliment the menu of pizza and popcorn, we came across *A Wrinkle in Time* (L'Engle, 1962). She had recently listened to the audio book and was amazed that there was a film adaptation. We snuggled up under our favorite blanket, which has inexplicably shrunk every girls' night over the past 5 years, and began watching. Although in my adolescence *A Wrinkle in Time* will always stand out as a book that awakened a sense of imaginative speculation about the world, I cannot say that I found the movie riveting despite its all-star cast. While I enjoyed watching my child feeling enthralled enough for us both, it wasn't long before I found my mind wandering as I sat cuddled under the blanket.

That's when the line hit me. The scene occurs when Calvin (a boy from school) sits down to dinner with Meg (the protagonist), Charles Wallace (Meg's younger wunderkind brother), and Meg and Charles' mom. Calvin is asking about Meg and Charles's father, who had disappeared four years prior. As Calvin inquires about Mr. Murry's work, the memory of Mr. Murry presenting a paper appears on screen. Mr. Murry says to the audience:

> So what if we are here for a reason? What if we are not just in the universe but the universe is within all of us? Think about quantum entanglement. Two electrons once bonded together, in love, if you will... suddenly separated by a galaxy...but somehow still just as connected. An unbroken union despite being galaxies apart.
>
> (*DuVernay, 2018*)

Prior to that moment, as I will explicate throughout this book, I had been struggling to describe the trajectories of affects. In my latest project, I had theorized how one might use magnetics as a metaphor for the way that hope is attracted and repelled through the intra-actions (Barad, 2007) of an assemblage. While magnetics offers fertile ground to discuss some trajectories, the binary of the metaphor felt constricting. I've never lived well with binaries, from the core of my racial, linguistic, and sexual identity to my scholarly work that has served to dismantle such understandings (e.g., Wozolek, 2018, 2019; Wozolek et al., 2017).

"Quantum entanglement?" I thought. "Of course. Quantum entanglement. How did I miss it?" It seemed so simple yet completely out of my intellectual reach until that line burrowed itself into my half-attending mind. However, as scholars like Gershon (2017), Hartman (1997), and Moten (2003), to name a few, remind us, there is movement even in silence and stillness. Although my mind was relatively at rest, the ideas still resonated across my ways of being, igniting new thoughts about my own work as it was entangled with my life. With that, the foundation of this project was conceived.

The point of this narrative is not to give too much weight to the almighty epiphany. There is, however, significance in acknowledging the deeply entwined nature of one's ontoepistemological understandings that are always in-action (Schwab, 1969). Would I have arrived at quantum entanglements without my daughter's perceived difficulty with reading, the fact that my son decided to go to bed early, or the brilliant invention that is girls' night? Maybe. Can one come to these understandings when things are unplanned, as this particular night happened to be? I'm not sure. However, as anthropologists, educational theorists, critical race feminists, and queer theorists alike have reminded us, we are always already inextricably all the parts of our vulnerable selves (e.g., Behar, 1997; Halberstam, 2011; Lorde, 1984; McDermott, 2011).

This book is a text of entanglements. The knotted parts of incalculable features that are currently becoming despite and because of dualities. It is a book that is intimately connected and, when finished, partially disconnected, from my own ways of being, knowing, and doing. Texts, as all writers know, take on their own life after they leave our desks and, as Gershon (2017) argues, are constantly in a process of being heard and unheard by readers. This is a book about beginnings that are entwined with ends that are also someplace in the middle, the moments of silence and noise that are felt in, through, and across contexts. This is a book about the always-already-not-yet—an assemblage.

References

Barad, K. (2007). *Meeting the universe halfway: Quantum physics and the entanglement of matter and meaning*. Durham, NC: Duke University Press.

Behar, R. (1997). *The vulnerable observer: Anthropology that breaks your heart*. Boston, MA: Beacon Press.

DuVernay, A. (2018). *A wrinkle in time* [Motion Picture]. Burbank, CA: Walt Disney Studios.

Gershon, W. S. (2017). *Sound curriculum: Sonic studies in educational theory, method & practice*. New York: Routledge.

Gottlieb, A. (2000). Rethinking female pollution: The beng case (Cote d'Ivoire). In J. S. DeLoache, & A. Gottlieb (Eds.), *A world of babies: Imagined childcare guides for seven societies* (pp. 115–138). Cambridge, UK: The University of Cambridge Press.

Halberstam, J. (2011). *The queer art of failure*. Durham, NC: Duke University Press.

Hartman, S. V. (1997). *Scenes of subjection: Terror, slavery, and self-making in nineteenth-century America*. New York: Oxford University Press.

L'Engle, M. (1962). *A wrinkle in time*. New York: Farrar, Straus, and Giroux.

Lorde, A. (1984). *Sister outsider: Essays and speeches by audre lorde*. Berkeley, CA: Crossing Press.

McDermott, M. (2011). Curriculum of the eye/I. *Journal of Curriculum Theorizing, 27(2),* 130–144.

Moten, F. (2003). *In the break: The aesthetics of the black radical tradition*. Minneapolis, MN: University of Minnesota Press.

Pinar, W. (1998). *The passionate mind of maxine greene: "I am–not yet"*. Bristol, PA: Falmer Press.

Pollock, D. (1999). *Telling bodies, performing births*. New York: Columbia University Press.

Schwab, J. J. (1969). The practical: A language for curriculum. *The School Review, 78*(1), 1–24.

Wozolek, B. (2018). Gaslighting queerness: GSAs, schooling, and Teachers' education. *Journal of LGBT Youth, 15(4)*, 319–338. doi: 10.1080/19361653.2018.1484839.

Wozolek, B. (2019). M/othering redoux: A Reading of curriculum theory to get through sleepless nights. *The Journal of Gender Studies, 28(8),* 873–882. doi: 10.1080/09589236.2018.1564649.

Wozolek, B., Wootton, L., & Demlow, A. (2017). The school-to-coffin pipeline: Queer youth, suicide and resilience of spirit. *Cultural Studies<>Critical Methodologies, 17(5*), 392–398.

अतिथिदेवो भव: ATITHI DEVO BHAVA

The content of this book can be rather jarring and triggering for some readers because the purpose is to think about the intricate nature of violence as it flows in, through, and across contexts. These entanglements of violence, or, as I have named them, *Assemblages of Violence* do not exist in an affective vacuum. They are always already knotted with affects such as hope and happiness, a point to which I will return several times throughout this book. Because the content can potentially be unsettling, it's worth taking a moment to welcome you to the context, the way one might welcome someone into their home, and ask them to ruminate briefly on some of the threads that are woven throughout the broader dialogue on violence.

Atithi devo bhava is a phrase in Sanskrit that translates to "the guest is close to God". It is a typical Indian greeting used to describe the relationship between guest and host. It is used here, on one hand, to welcome the reader and, on the other, to describe the relationship between the narratives and theories rooted in these stories. This book tracks back and forth between two research studies—one in India with women who were survivors of domestic violence and one with queer youth of color at an East Coast middle school. While the layers of each study will be unfolded throughout the book, this chapter is meant to position myself within each context and then to give some insight into the narratives themselves. Toward this end, this chapter is divided into four sections and the first section begins with my childhood in Ohio to explicate questions of race and queerness that were central to my youth. The second section focuses on a brief description of the middle school where the first study took place. The third moves back to my personal experiences in India, as the daughter of an immigrant father who did not have the privilege of visiting the country often in my youth. Finally, this chapter will illustrate some of the context behind the women's shelter where the second study took place. The collage that is constructed through these sections is meant to give an invitation and welcome to my intellectual home, a chance for the reader to know the place from which these theories were born.

Unsnarling Chains: Contextualizing the Researcher

I was about eight years old when my family moved from central to northeast Ohio. As my parents sorted through boxes, one of the chores that my mother had wisely delegated to keep me occupied was to organize her jewelry box. Until this moment, my mother's jewelry had remained largely a mystery, as I was consistently instructed not to play with her necklaces and earrings. At first, I was elated at this task because it meant that I was finally permitted not only to open but also to touch the jewelry. Little did I know that the contents had shifted during the move, leaving behind an evil mess of objects that, in my current memory, seems worse than what the fate found upon opening Pandora's Box.

Ignoring the impending frustration of the challenge, I sat down excitedly and began to try to unravel the chains. If you've ever worked on a knot, you understand the nested emotions that came next: my feelings settled somewhere in the middle of angry, frustrated, hopeful, and determineds. The more one side loosened, the more cinched another would become. The experience was met with tears, laughter, celebrations, and a few hopeless, Muppet-like screams. I would like to say that I unsnarled every necklace, but the reality is that if I were to visit my mother's house today, I would likely still find a few tangled chains, forever bonded in their knotted configuration.

A few months into living in this new town, some boys from the neighborhood asked to play cops and robbers with them. Having few friends, I happily agreed to the game. Within minutes, I found myself chained to a tree while they interrogated me about my father's green-card status. I did not know what a green card was and, instead of answering them, decided to vehemently insist that I did have a yellow library card. I explained that I could get a green one if they would just let me go. They did not.

I sat chained to the tree for some time, recalling the knots in my mother's necklaces, afraid that if I moved too much in one direction, some other part of me would become more trapped. The metal on my body reminded me of the cold metal between my fingertips as I gently worked to free each piece of jewelry. It was not until the sun was setting that one of the boys returned to release me, provided I did not tell anyone about the ordeal. Following his instructions, it was not until my thirties that I finally confessed the event to my mother. I would like to say that I've unsnarled myself from the chains of the "game", but when my daughter wants to visit someone on a playdate, I can still feel the knotted parts on myself that are traumatically, and likely forever, impacted by the experience.

The East Coast, Research Context One: Defining the "Queer Problem"

The glass Coke bottle spun almost out of control until halting suddenly at Alicia,[1] a slender brunette girl that I knew from my social studies class. There was a palpable silence; this was the third time the bottle had landed on this girl when I spun it. "Are the bottle and carpet fibers conspiring against me?" I wondered as I studied her perfectly smooth, oval face. Her dark, almond-shaped eyes were wide, and her full lips were slightly parted with confusion at the results. After all, what were the odds? After a moment of silence, the crowd of hormone-laden eighth-grade teenagers gave a loud holler, complete with overly dramatic smooching noises.

> "You know what that means, right?" the party hostess declared, "seven minutes in heaven!"

When a bottle landed multiple times between two people, "seven minutes in heaven" was a punishment or reward, depending on one's position on being pressured into publicly kissing and, of course, on the other person involved. The lucky "couple" was to go into the nearest closet and kiss for seven minutes while their peers eagerly listened and speculated outside the door. My heart stopped. "Was this really going to happen?" I wondered. Before I could protest or lean into the situation, Alicia objected loudly, shouting, "Only if I'm paid!" A boy reached for his wallet and offered ten dollars. Everyone laughed and the conversation, as well as the bottle, moved on.

I've thought about that moment often as one of the first times I publicly grappled with questions of sexuality. Like many young people, I had plenty of experience thinking through sexuality in private, wondering how and when I might find answers to the secret questions that I buried more deeply than any other insecurity I possessed at the time. After the first part of the party, the boys left and the girls stayed for a sleepover. We sat around and gossiped about who everyone had kissed, how it felt, and if they would do it again given the opportunity. I remember slowly spinning the bottle on the ground in front of me, half listening to the banter while reflecting on the events. The bottle was cold and hard, almost enough to center me as laughter from the other guests ebbed and flowed.

> "How about you, Boni?" asked the girl with whom I was almost pulled into heaven. "Who did you like kissing?"

I laughed, nervously spinning the bottle one more time, and started to describe kissing Sean, a popular boy who was in a band. My detailed account of the kiss seemed to satisfy everyone's curiosity and I returned to spinning the bottle quietly on the floor, wondering if anyone else was left so introspective by the game. Over time this "almost" moment has slightly faded from my mind, yet its edges remain sharp in my being when I recall it. I did not speak about the game again until years later in an intimate conversation with a woman in college, but when I saw glass bottles—which became less and less available as the plastic industry plagued the planet—I couldn't help but think of my lost and exposed feelings that night.

As a postdoctoral research fellow at Loyola University Maryland, I found the memory of this event running through my mind as I set up my office. Prior to completing my doctoral program, I had run Genders and Sexualities Alliances as a K-12 teacher in the Midwest. While my coadvisor and I had managed to build a rather strong program, I recognized the significance of keeping my own positionality quiet because of the intensely conservative nature of the school and its administration. As I decorated my new space, I wondered how many of my experiences, both personal and professional, I should bring in to work at this small Jesuit institution. This was, after all, a fresh start.

One afternoon I was setting up my office and deciding just how many rainbow flags were appropriate for a professional space, when a colleague knocked my door. She had received a request from a school that she wanted to discuss with me. The call had come from New Horizons Middle School, a small charter school located in the city. Charter schools in this East Coast state worked differently than others across the United States because, at the time, charters were public schools. This meant that they were funded and governed by local Boards of Education, but operated independently from central administration (Broadwater, 2019). This particular school had supplemental income from a local business and, like many charter schools in the city, its independence from central administration was a point of contention between it and traditional public schools. This autonomy also meant that the administration at New Horizons operated without much consideration for intradistrict implications, something that became rather relevant as I spent time observing the school as a researcher.

My colleague explained that some students at New Horizons had come out to teachers and, without the tools to negotiate the students' brave choice to be out, the teachers followed what they felt were proper safety protocols: they called the students' parents or guardians. While the teachers were aiming to help, their decision to inform parents about students' sexual orientation or gender identity was based on a misunderstanding about their legal responsibility surrounding what needs to be reported to parents. Because no students were recorded as at risk to harm themselves or others, it was not these teachers' legal responsibility to report such information. As a result of these calls to home, some students were kicked out of their homes while others found unexpected support from their families. Recognizing the error of her colleagues, one of the school teachers, a former student of my institution, decided to reach out to my university for help. My colleague received this call and came to me because, as she said, "You're the only one who does that kind of work here".

Since my scholarship focuses on questions of race and queerness, I was not hired to intervene in public school issues. However, given the dearth of scholars whose lines of inquiry focused on queerness and race at this institution, my colleague's options were limited. After getting in touch with the teacher, Bridget, who had contacted my institution, I arranged a trip to the school. Although centrally situated in the city, New Horizons was located beside the harbor. As waves slapped on the wooden barriers, the school felt rather secluded from the rest of the bustling downtown. The 162 sixth-to-eighth-grade students who attended the school were selected through an application process facilitated by the city's lottery system. Students were predominately African American and Latinx.[2] While gender identities and expressions were in flux for some students, New Horizons reported an even split of the genders assigned to students at birth.

After being permitted into the main gate of the school, I made my way across the gravel parking lot and was buzzed into a small, wood-paneled blue building. I signed in and was greeted by Bridget, a young, tall, English Language Arts teacher. Once in her classroom, we settled down at a low table as we waited for Joel, the school's principal, to arrive. The room, which normally housed 25 sixth graders, had all the controlled chaos of a classroom at the beginning of the school year, complete with papers thrown rather than neatly stacked in their assigned places and books hastily tossed on to the shelves.

When the principal, Joel—a thin, thirty-something, dark-haired man—strode in with an air of extreme confidence, it became clear in a matter of minutes that he was not interested in professional development. What he wanted was someone to help him solve what he called the "gay problem", meaning that he wanted a professor to meet with the students to "give them guidance" about their developing gender identities and sexual orientations. About 20 minutes into the conversation, Joel looked at me and asked, "Talk to me about your experience with queer youth. I mean, what do you get out of doing this? Why are you *really* interested in this kind of work?" I could feel the desk, cold and hard under my hands. It reminded me of holding that bottle as it grounded me between responses, keeping me firmly in the moment while my mind raced. Queerness, as scholars such as Halberstam (2005), Hall and Jagose (2012 (2012), Gilbert (2014), and Johnson (2016) have argued, is easily suppressed in such moments of uncertainty. I thought of my office, the flags that decorated my whiteboard, and then back to that game of spin the bottle. How much of yourself do you reveal in these delicate moments? How much do you suppress?

> "What I get out of it", I began slowly, "is the ability to research, should you decide you want me here for that purpose. What I get out of it will hopefully be publications. But mostly, what I get is knowing that I can think with youth about their experiences, knowing I did not have anyone thinking with me through mine at their age".

He sat back and smiled. "Good", he said. "We will get a list together of kids for this group. Bridget, can you do that?"

"A list?" I asked, still feeling off kilter from his response, which seemed to denote a rather simplistic conceptualization of who is qualified to work with queer youth. Once I perceivably checked this box for him, he seemed content to continue with the project.

> "Yes", Bridget jumped in. "We want you to have a list of strong kids, a medium-sized group, nothing too large or difficult to handle but not so few that they will get bored".

"No one who is going to be, well, too embarrassing", Joel interrupted, pointing his finger toward the ceiling as if to punctuate his point.

"Well, you know, we have kids who can be a challenge, behaviorally speaking", Bridget clarified. "We don't want those kids. We want kids who will be…a good fit. We will send out a survey, but we will ultimately decide who belongs in this group, okay?" While my gut response was to interrupt this kind of sorting, I decided to let it play out a bit and then think through the process with Bridget and Joel at a later date, especially as the survey was rather innocuous and not immediately harmful to students. As qualitative researchers have long argued, the line between observer and participant is often a difficult one to traverse, particularly with the ethical considerations in play for marginalized youth.

As a result, the initial list of students I received for the group were not necessarily those who had asked to be a part of an extracurricular activity but, rather, those that teachers had identified as potentially queer. Much like Geertz's (1973) anecdote about turtles all the way down, it became clear that this school had problems—all the way down. While the decision to identify kids based on perceived queerness was upsetting after my previous work in the Midwest, where I had witnessed what Sedgwick (1993) refers to as the indelible reality of queer youth suicide, I was infuriated but not surprised. This is where my study began—with a list of students who made up "the queer problem" at New Horizons, and flashes of doubt about how to think with students about their own coke-bottle experiences.

India, Research Context Two: Getting to the Petticoat of Things

I remember the first time my mother dressed me in a Sari. I was raised in a small rural town in the United States and the inconspicuously placed bin under my parents' bed that held my mother's saris was a mystery to me. Every so often she would open it up and the smell of silk and metal tinged with mothballs and sandalwood would fill the air. I would run my hand across the thickly folded material, wondering what pieces of my father's culture I was missing by living oceans apart at a time when the internet was relatively nonexistent and international phone calls were rather expensive. Although my brother and I were raised in the Catholic Church, Indian society was predominantly Hindu. In search of Indian community, our misplaced-looking multiracial family, with children who knew little to nothing about Hinduism, would occasionally attend Diwali and other celebrations. For me, these events signified a space where, although the color of my skin felt as exotic as it did in my predominantly white, rural elementary school, my soul felt at ease surrounded by families who were also likely perceived as "different" in Ohio. My heart felt lighter watching my father interact with other Indians, as it was one of the few moments that his dark skin and ways of being "belonged" in any space. I was always amazed at how quickly the accent he often denied himself at work, accompanied by the unmistakable bobble of his head during conversations, came flooding back upon entering the room.

The first time I went to India was when I was twelve. The flight seemed to infinitely drag on as my brother and I drifted between boredom and exhaustion while we watched Bollywood movies on the screen at the front of the plane. We were not raised speaking Hindi and, as I stared blankly at a woman on the screen singing joyfully, yet inexplicably, in the rain near a tree, my first period started. Needless to say, our arrival in my grandmother's village was braided with my excitement at being in a place that had existed only in my dreams, meeting family for the first time, and wrestling with my newly experienced menstruation.

The morning of our trip from my grandmother's village to the mountains where my mother had stayed as an exchange student decades earlier, my mother placed a sari with matching petticoat and blouse on the bed. She explained how to wear the sari correctly, its significance as it is correlated with puberty and marriage, and the importance of making sure that the petticoat remained unseen. She then carefully wrapped my developing body with a sari from the bin that I had eyed with envy too many times. It was bright pink and lime green with a Gaudí-esque tree pattern that crept up from the bottom of the silk. As proud as I felt in the moment to be marked as a young woman and, as I would later learn, to be accepted into the fold of kitchen conversations with other women as a result of my newfound maturity, I was confused by the petticoat. I kept wondering why what I perceived as a skirt was meant to be kept private. Yet, as my grandfather sauntered by the room, I felt naked standing in my petticoat and blouse, waiting for my mother to wrap me. While she folded the fabric gently around my hips, I couldn't help but feel as though the most significant piece of the outfit was the petticoat. After all, it held everything else together. No petticoat, no neatly wrapped Sari.

As I watched the dynamics within my own and other Indian families, I began to understand petticoats as a metaphor for the lives of many of the women I observed living in traditional lifestyles. They often went unrecognized for their work and yet they supported and maintained the folds of family dynamics. Although the petticoat goes unseen, it offers shape and definition to the sari. The women similarly existed, quietly shaping their contexts within the constraints of the patriarchal fabric of their lives. The petticoat can also be understood as carrying a degree of coloniality, a way of hiding the body of the "other" that was all too often sexualized under colonial rule.

Eventually, as an adult, I wrestled with my concerns and experiences with violence against women and girls. Struggling with the reality of living with/in such acts of aggression, I often found myself recalling this childhood metaphor of the petticoat. I wondered how women in India struggled with, and enacted agency against, violence that endured from fathers, husbands, and sons alike. While I had heard stories told in hushed tones in kitchens as I stared at the Gaudí trees creeping up my sari, at holiday events back in the United States, and in subsequent visits as a young adult, I had not participated enough in the culture to understand what daily survival entailed. Reflecting on my dissertation data that included questions of sexual assault and abuse of the young women who participated in my study, I began to wonder how acts of violence transcended place. It was as a postgraduate that I decided to reach out to a women's shelter in India to further investigate these questions.[3]

Getting in touch with a shelter while living in the United States proved to be rather difficult and, as a result, I reached out to contacts in India to help. After several months, I was given the name and phone number for Aasha Women's Shelter but was only able to reach the director by phone once prior to my trip. Although during that call I had arranged a visit and permission to conduct research, there was an understandable amount of mistrust from the director. As a result, she decided to withhold the exact location of the shelter until I arrived in the country. Instead, I was told that a driver would be arranged once I had settled in after my arrival.

Upon entering the country my contact, Sevita, was suddenly unsure if we would be able to visit the shelter. She said she had to wait to receive a call from the driver and the shelter. Finally, after several days, I was informed that we would leave in the afternoon. We drove for about an hour into a smaller village and pulled over at a bookstore. The driver got out and had a conversation with someone inside the store, and we were then motioned up a hill where we were told that we would find the Aasha Women's Shelter.

After this secrecy, the shelter was surprisingly visible. It was marked with a large sign and protected by a metal gate and a small pack of dogs that barked excitedly upon our arrival. Cows sauntered by the driveway while women and men carried baskets on their heads as they worked their way up the hill from the morning market. I was met by the sounds of children playing on a schoolyard nearby, seeming to signal that life continued regardless of the circumstances that brought women to this place. Sevita and I made our way up the stone path and through the large yard that sprawled in front of the white, concrete building. We were greeted by two middle-aged women who sat behind a wooden desk in a small room decorated with paintings and woven fabrics. The room smelled heavily of earth and mothballs. Although the room's design allowed the wind to cool this tiny vestibule rather effectively, an old fan still whirred frantically in the corner.

Neetu and Deepti, the center directors, sat with their arms crossed, looking like aunties who had just discovered that I had ignored their advice and were looking to, at best, have a stern word about my behavior. I must have approached with some hesitation because Sevita grabbed my elbow and, with a disapproving click of her tongue, pulled me toward the women at the desk.

"Hello!" she announced with a soft, friendly tone. "I am Sevita and this is Dr. Wozolek. We spoke on the phone about her talking with you about this place". Both women simultaneously shifted their gaze from Sevita to me, waiting for me to use my own voice.

"Yes... I'm Dr. Wozolek. We spoke about three months ago", I managed to explain.

> "Yes. That was me", Neetu explained, peering at me from behind her thick glasses. I glanced at her face. Her gray hair was neatly pinned up, almost sparkling under the fluorescent lights. Although she seemed to be in her fifties, her high cheekbones gave her a youthful appearance. Her voice was low and commanding as she looked at me and asked, "But why are you here, my dear?"

"Well… As I told you on the phone…" I began.

"No. I mean, why did you *really* come here?" she asked.

As researchers, part of our job is to be transparent about our interests. This is important for everything from Institutional Review Board approval to a clearly articulated line of inquiry for tenure and promotion. Although such matters may seem to be and often are reduced to bureaucratic hoops, they are nonetheless important acts of ethics. We need to be clear with participants about our attraction to a site. While there are many professional reasons to think with people at particular research sites, the reality is that all too often we chase the answers of our own deeply seated troubles (Behar, 1997; Rosaldo, 1993). It was because of this last point—the experiences that had kept me up at night but rarely found their way onto the pages of my work (Mahmood, 2008)—that, despite having prepared an answer in advance that spoke to my life and expertise as an academic, my mind went blank. Instead, I began with what I knew in the moment: "There's so much violence against women. I don't know where to start. I guess thought I'd start here". I swallowed hard, feeling as exposed as standing in my petticoat, waiting for my mother to wrap me. Both eyes of Neetu and Deepti softened and, without a word between them, they collectively exhaled.

Looking back, I'm sure this work started sometime between listening to women talk about abuse in the kitchen as I wondered if those surreal trees would grow up my sari and shackle me with their branches, and the moment I decided to be vulnerable with Neetu and Deepti. As Behar (1997) argues:

> loss, mourning, the longing for memory, the desire to enter into the world around you and having no idea how to do it, the fear of observing too coldly or too distractedly or too raggedly, the rage of cowardice, the insight that is always arriving late, as defiant hindsight, a sense of the utter uselessness of writing anything and yet the burning desire to write something, are the stopping places along the way.
>
> (*p. 3*)

Regardless of how I arrived, there I sat, feeling vulnerable in an old, hard wooden chair. I found myself feeling distant from my own narrative while simultaneously struggling for emotional distance from the narratives shared at Aasha Women's Shelter. Perhaps this is why, as Behar might argue, research can both break and heal the soul.

Conclusion: At the Intersection of Contexts

Aside from personal catharsis, my time at both New Horizons and Aasha Women's Shelter deepened questions about violence that had been percolating over the course of my life and across personal and professional contexts. From a global perspective, life in the United States is punctuated by violence against Black and brown people, queer-identifying communities, and women. As the Black Lives Matter movement has increasingly become a global phenomenon, with protests held in the United Kingdom, Germany, Iran, and across Latin America, considering how race-based violence propagates is increasingly necessary. As trans women of color continue to be maimed and killed across the United States and throughout the world, careful consideration of how violence moves has become paramount. As women across the globe continue to deal with intimate partner violence and other forms of assault, spurring the #MeToo movement and engendering other spaces of help, tracing and interrupting the normalization of aggression is incredibly significant.

Returning to Behar's (1997) quote above, while many countries are wrestling with questions of equity and access that are denied within historical and contemporary norms, perhaps we have all "arrived too late" as the oppressed have already been, and continue to be, hurt by such norms. While I consider what it means to unsnarl the knots created by, for, and in the wake of violence (Sharpe, 2016), I have to hope that, although late, this book may yet serve as a "stopping point along the way" to interrupting everyday violence in schools and across communities. By approaching acts of aggression with vulnerability, I hope to offer a way to heal what has been broken by assemblages of violence.

Notes

1 All proper nouns have been replaced with pseudonyms to protect the privacy of participants throughout *Violent Assemblages in Education*.
2 While current trends use "Latinx" across the United States, this construction is still fairly Western in its use and Latin@ remains an international construction for gender inclusivity. However, students in the study used "Latinx" in writing to describe themselves and, therefore, Latinx will be used throughout this text.
3 My use of national rather than state identification of this shelter is intentional. Although this shelter is well-known in the town where it is located, it was not well advertised within the region for safety concerns. For this reason, the exact location of the shelter will not be disclosed.

References

Behar, R. (1997). *The vulnerable observer: Anthropology that breaks your heart*. Boston, MA: Beacon Press.

Broadwater, L. (2019). Governor Larry Hogan announces plan to boost charter schools in Maryland, *The Baltimore Sun*. Retrieved from: https://www.baltimoresun.com/education/bs-md-hogan-charter-schools-20190123-story.html.

Geertz, C. (1973). *The interpretations of cultures*. New York: Basic Books.

Gilbert, J. (2014). *Sexuality in school: The limits of education*. Minneapolis, MN: University of Minnesota Press.

Halberstam, J. (2005). *In a queer time and place: Transgender bodies, subcultural lives*. New York: New York University Press.

Hall, D. E., & Jagose, A. (Eds.) (2012). *The Routledge queer studies reader*. New York: Routledge.

Johnson, E. P. (Ed.) (2016). *No tea, no shade: New writings in black queer studies*. Durham, NC: Duke University Press.

Rosaldo, R. (1993). *Culture & truth: The remaking of social analysis*. Boston, MA: Beacon Press.

Sedgwick, E. K. (1993). The epistemology of the closet. In H. Abelove, M. A. Barale, & D. M. Halperin (Eds.), *The lesbian and gay studies reader* (pp. 45–62). New York: Routledge.

Mahmood, C. K. (2008). Anthropology from the bones: A memoir of fieldwork, survival, and commitment. *Anthropology and Humanism*, *33*(2), 1–11.

Sharpe, C. (2016). *In the wake: On blackness and being*. Durham, NC: Duke University Press.

INTRODUCTION

Violence is complicated. One only needs to recall the violent processes that transpire when a person has an orgasm, gives birth, or dies to consider the contexts and complexities that are central to a conversation about violence. The narratives presented in the Atithi Devo Bhava are certainly fraught with violence that is complicated by ordinary and extraordinary affects and their imbricated contexts. While there are many inroads to such dialogues, here I attend to sexual relationships as a reminder that intention, attention, and expression (Gershon, 2018) matter. This is because, although the primary focus of this book is not sex, questions of consent and attention to sexual orientations are central across the data presented in this text. The book theorizes how violence is made, unmade, and remade through assemblages and how it always exists in relation to other affects, like joy. In terms of sex, consensual relations can be violent in their expression (see, for example, BDSM (Bondage and Discipline (BD), Dominance and Submission (DS), Sadism and Masochism (SM)) culture) but not in the body's interpretation of pleasure as it is related to that expression. In contrast, nonconsensual sexual relations might have an expression that is perceived by one party as loving but felt by the other as violent. Violence is therefore as much about the intentions of the bodies involved as it is about the expression of affect and action. This is not to say that certain kinds of violence are not clear across layers of scale but, rather, that I pay careful attention to the multiplicity of spaces where violence ebbs and flows.[1]

In this book, I conceptualize violence not only as a complex set of affective interactions but also as agential in how it impacts bodies. Although affect is often discussed in terms of movement between exclusively human bodies (Gregg & Seigworth, 2010, 2010), I conceptualize affect both as it travels in, between, and through human bodies and as a set of complex intentions between human and nonhuman bodies. This is significant as nonhuman bodies can also carry affect, a point I return to in Chapter 1. Considering the movement of affect is done with attention to questions of direction and the dynamics of expression that erupt from such movement between bodies. Tracing the trajectories of violence is therefore an incredibly entangled process. Think, for example, of the times that feelings of joy precede conversation or when walking into a room feels "off", even without prior knowledge about the context. As I argue throughout this book, the value in conceptualizing affect as agential is that it shifts our understanding from a force that sticks between bodies (Ahmed, 2010) to one that can also entangle with bodies through intra-action (Barad, 2007).

The next section delves into thinking about how violence moves with agency. In complexity-that-is-violence, it is important to think about how violence moves; in examining these complexities, one can also think about how they might be interrupted. In this case, I think through such movements within a framework of sonic understandings (e.g., Gershon, 2017; Goodman, 2012; LaBelle, 2018; Sterne, 2012, 2012) to explain what Barad (2007) discusses as intra-action. Barad argues that bodies—both human and nonhuman—do not just *inter*act but, rather, they *intra*-act. For the purpose of this book, a "body" can be understood as anything that can be agential. This means that everyday affects like hope and joy or forms of oppression like racism can be understood as "bodies", a point to which I return in Chapter 1. As sounds break silences, other noises, and bodies (Goodman, 2012; Miller, 2005; Moten, 2003), they also reverberate across contexts.

I understand these reverberations through what Barad (2007) discusses as *intra*- rather than *inter*actions. Barad uses intra-action to describe the ways that social, political, historical, and contemporary bodies come together in an event to produce a coconstitutive agency among the bodies involved. The difference between inter- and intra-action, Barad argues, is that in interactions the subjects may engage in an event and remain unchanged. In intra-actions, subjectivities are changed and produced through the event. As I argue below, Barad's understanding of intra-action is not unlike the friction and force with which we experience sound. Within the event, Barad argues, bodies are shifted. Further, Barad is clear that these intra-actions are composed of both human and nonhuman actors; the following section uses Bach's Cello Suites as a primary example of such movement. The Suites resonate within the music itself—the original composition as it meets the contemporary cellist, intersecting and entangling with the audience. Violence similarly intra-acts in polyphonic and heteroglossic (Bakhtin, 1981) phrases.

Bach's Assemblage

Central to the nested and knotted nature (Nespor, 1997) of violence across contexts are the ideas of intention, attention, and expression. In his discussion of resonance and reverberation in research and sound arts, Gershon (2018) argues that:

Intention can be understood to lead to notions of attention across multiple layers between and within producers... Attention pertains to what and how producers went about with their intentions...Expression is the outward release of something that is a result of producers' processes of intention and attention. (p. 2)

Take the étude-like nature of Bach's Cello Suites, a piece that can easily capture the imagination of listeners. Although performed by one instrument, Bach's work conjures an entire world of instrumental possibility, invoking intensities that exist across states of being. It is simultaneously life affirming and joyous, yet grief stricken and melancholy. In these and other ways, Bach's Cello Suites themselves move, much as they set in motion other bodies with a rhythm that is grounded in both the material and the abstract. The stunning composition of the suites calls the listener within and through states of being, knowing, and doing (Gould & Keaton, 2000), resonating across bodies as is leaks from what some perceive solely as sonic possibilities—the music as it is heard—into more sensuous understandings—ontoepistemologies borne across the senses, regardless of their origin within the sensorium (Pink, 2007; Rosaldo, 1993; Stoller, 1997). In other words, it's not just about what one hears nor, ultimately, what is misheard (Gershon, 2018), but rather about the way that the suites resonate and reverberate across entwined contexts within and outside of the listener.

There are two significant considerations regarding the Cello Suites and Bach's work in general. First, to date, little evidence exists in the form of Bach's personal reflections or sketches on the suites. Additionally, none of Bach's original manuscripts of the Cello Suites have been recovered. While this has not stopped many from conjecture, Bach's actual intent behind the creation of the suites remains debated among enthusiasts and scholars alike. However, it is known that Bach wrote with pedagogy in mind (Spitta, 1951; Winold, 2007). Although Bach himself received relatively few years of formal education, he often articulated the significance of using his position as a composer to teach ideas that ranged from religious instruction to music theory for students (Bettmann, 1986; Siblin, 2010). Toward these ends, Bach's expression of his ideas and ideals was often a complex set of interrelated concepts, as heard in the Cello Suites.

Second, the Cello Suites can be understood to express the power of resonances and reverberations, raising two important points. On one hand, there are resonances within the music that are exquisite in their complexity. For example, Bylsma (1992) argues that this composition poses the question: "How many notes can one take away and still leave a complete suite in the mind of the listener, harmony and counterpoint included—not forgetting the many dissonances and syncopations which exist only in the memory of the audience?" (p. 5). Bach relies heavily on the listener to complete phrases that he has intentionally left incomplete while simultaneously immersing the listener in a polyphony of resonances created by notes left ringing well after the cellist has moved on (Waltham-Smith, 2006). The complexity of the suites exists within its intrinsic dynamic resonances (Boon et al., 1990) as they reverberate extrinsically between bodies.

On the other hand, the resonances and reverberations are not contained to the score and, upon spilling intentionally into the audience, impact the listener. Music is as much an aural experience as it is a haptic one (Altenmüller, 2004; Peretz & Coltheart, 2003). It is as much grounded in the moment as it is in broader sociocultural ideas and ideals (Feld, 1995; Love, 2016). Gershon (2018) reminds us that resonance and reverberations are at once the way ideas, sounds, and movements vibrate in and through a body and are also the "never-ending ebb, flow, and interactions of any given thing in relation to ecologies" (p. 2). Conceptualizing the complex interrelated ways that bodies affect and are affected, Gershon argues that attending to resonances and reverberations is significant to giving the "dignity of attention to [why something] resonates with that person at that time and… the reasons [that person] believes the resonance to be important" (p. 2). Treating thoughts with dignity, Gershon argues, does not exclude a person's right to critique, disagree, or reject something. Perhaps more importantly, he suggests that taking an idea seriously, no matter how oppressive or absurd it may seem, is central to forming a strong dissent.

When an idea resonates with one body, it is possible for them to become a vehicle through which the reverberations of that idea are carried. For example, in 2019 the US-Mexico border crisis under the Trump administration left children traumatized as they were separated from their families, devastating refugees as they were deported and turned away from asylum. Amid the national dialogue, world-renowned cellist Yo-Yo Ma gave a concert at the border where he played Bach's Cello Suite No. 1. In an attempt to connect cultures through The Bach Project, Ma used music as a disruption to oppressive structures, urging the audience to use culture to "build bridges, not walls" (Ma, as heard on *All Things Considered,* 2019). Within this political articulation of the suites, Ma intended for music to be a space of reflection, dialogue, and healing in a place of unrest. Using the Cello Suites, Ma exemplified what can happen when a sound is carried with intention and attention by and through people.

The multiple movements of the Cello Suites can be thought of as polyphonic, written across lines that converge into the suites' musical texture. Bach's music operates as heteroglossic in that it never utters singular lines and the music metaphorically "speaks" across spaces and places. The suites are therefore inherently heteroglossic, forming worlds within words (Bakhtin, 1981) as the text of the music is read, heard, and felt across contexts. Any one utterance of the suites contains a multiplicity of other sounds as well as sociopolitical, cultural, historical, and contemporary ideas. Everything is in relation, even when the relations are carried by one voice, in this case by a single cello. The Cello Suites stand as but one example of these complex understandings of sound, any sound could serve as an example of these complexities. Just as any one utterance can be understood as heteroglossia, any sound can be an example of polyphony. Whether it be a hip-hop artist spitting verses, a Gregorian chant, a wave crashing in the ocean, or a child laughing, all sounds are entangled across spaces and places.

Finally, it is important to remember that making any sound involves friction (Tsing, 2005) or force. Whether it's the bow on a string, air moving past the larynx, or a hand on a drum, friction and force are central to the process of producing noise and, subsequently, to being heard. Although mechanically different, friction and force impact all actors imbricated in and across sounds. Tsing (2005) describes friction as a space of productivity and destruction: without friction, there is no movement. However, such productivity is tempered by the idea that friction triturates bodies as much as it moves them. Just as a tire is worn down by friction, so is the road over time. The tire's ability to move is therefore shaped by present and past intra-actions with friction and force.

Gaslighting as Violent Intra-action

Violence exists within the same complex framework as sound. Consider for a moment the power of psychologically manipulating someone until they question their own sanity, or, as it is colloquially known, gaslighting them. Just as Bach's Cello Suites are a strong example of a polyphony and heteroglossia, so is gaslighting a key example of polyphony and heteroglossia. Gaslighting is but one form of covert control that is employed in multiple contexts (Sweet, 2019), from gender-based abuse (Harris, 2018), to homo- and transphobias (Riggs & Bartholomaeus, 2018), to manipulating the elderly (Sweet, 2019), to gaslighting based on race (Tobias & Joseph, 2018). Gaslighting exists in schools (Wozolek, 2018) and occurs in intimate partner relationships at home (McCauley et al., 2018). From workplaces to social media (Fuchman, 2019), gaslighting is quite present in and across societies.

What makes this practice so insidious is that the person being gaslit is left intentionally questioning the events, positioning the victim in an echo chamber of hard-to-name abuses. Often, because gaslighting can feel so nebulous, the victim may express hesitation in describing the events as abuse. This gives the aggressor more room to deny the dehumanization perpetrated through gaslighting, which serves to continue the cycle of abuse. Constructs such as racial battle fatigue (Smith et al., 2007) and social movements like #MeToo provide powerful examples of how these cycles of violence are enacted in everyday experiences. Victims are often left wondering if the speaker was being intentional in their harassment, if the intent was directed specifically toward the victim, or if the victim is reading too much into the verbal assault.

Successfully engineering gaslighting is a tool wielded by those with power to engender and maintain power differentials. While gaslighting might not have the same intention as a piece of music—although some music is certainly composed with violence in its intention and attention (e.g., Futrell et al., 2006; Levi, 1990)—it does have similar complexities in its entanglements. First, there is the physical impact of words on the body. Sounds travel through solid, liquid, and gas. While the decibel of an insult might not be enough to break a glass per se, it does have a physical impact. There is a reason that sonic warfare (Goodman, 2012) is a developing military science, one that has leaked into schools and prisons as a form of control (Winberg, 2019). However, as almost any person who has been the victim of gaslighting and verbal abuse knows, sometimes just a person's tone can resonate in ways that raise the alarm about safety. This is at once a literal crashing of sound waves on the body as well as the collision of one's histories with the next moment of aggression.

The relations between bodies often come into sharp relief during moments of violence. Describing an instance of domestic violence that occurred as a man was watching the Super Bowl, Massumi (2002) argues that it is not just the relationship between the couple that might be fraught with their interpersonal history of violence. Moments of violence like this are also driven by factors like the television, masculine culture as it relates to football in the United States, and social and interpersonal codes that were upset by the game itself. These intra-actions occur in the act of gaslighting as well. In gaslighting, both the aggressor and victim are knotted within sociocultural norms and values and further imbricated with their own personal and collective histories. Instances of gaslighting therefore never exist in a vacuum and are always imbricated with other bodies.

The relationships between actors—the survivor, her partner, and football—that Massumi (2002) pulls through his example are strong illustrations of how human bodies are entwined with nonhuman actors. Barad (2007) similarly discusses the entanglements between bodies to reframe questions of agency. Other examples might be how the listener, the Cello Suites, and political forces are in relationship; how the students at New Horizons were knotted with systems of schooling and broader ideas about queerness; or how women's roles are imbricated with instances of violence. Throughout this book, bodies and what constitutes a body are centered along with dialogues about what bodies potentially produce, reify, and deconstruct through their relations.

Finally, the concert of gaslighting often braids together the aggressor, the gaslit, and those who witness the abuse. It should be noted that, as Hartman (1997) argues, there is a fine line between a witness and a spectator, though both are entwined with the event itself. Unlike the spectator, the witness works to "give expression to outrages without exacerbating the indifference to suffering that is the consequence of the benumbing spectacle" (Hartman, 1997, p. 4). Moten (2003) builds on Hartman's dialogue by noting that reproduction is inherent to both the witness and the spectator, though what is reproduced is a question of intention and attention.

This framing is not unlike the relationship between the music, the cellist, and the audience. It is also similar to the women in India, their partners, and those (like family) who witness aggression. Returning to Bach's suites, when the audience moves the music across contexts, as did Yo-Yo Ma before he mastered this particular composition, the witness becomes enmeshed in knotted roles: at once a witness and a musician. Ma's expression of these roles pulls together potential musicians, practicing performers, and those who are not musicians but nevertheless feel music. In witnessing, others' intentions might be shifted as they are folded into the event. Gaslighting works similarly, in that it is often interrupted when someone other than the person being gaslit becomes the audience or a witness. One only needs to recall how watching aggression, like homo- or transphobia, can normalize other forms of violence to understand how those outside of the immediate context are still intimately connected with the event. This speaks not only to the complexities of gaslighting but also to entanglements and assemblages as they are populated with bodies and intra-actions.

Violence *is* complicated. It is impossibly intertwined in ways that make it difficult to trace its trajectories. Given the inherent complexities of the many forms of violence, this book envisions all violent intra-actions through an assemblage. By "assemblage", I am referring to the messy and entangled intermingling of bodies, both human and nonhuman, that are productive in that they give "expression to realities, thoughts, bodies, affects, spaces, actions, ideas, and so on" (Weheliye, 2014, p. 47).

Specifically, the purpose of this text is to think about assemblages of violence (Wozolek, 2018). Discussing violence as an assemblage importantly recasts the singularity through which violent interactions are often discussed. Instead, the idea of assemblage insists that violence is iterative, never singular, and is always necessarily dependent on sociocultural norms, histories, and other acts of violence. In the case of youth who die by suicide, for example, mental health professionals often discuss how the moment of self-harm is always precipitated by a multiplicity of events. *Assemblages of Violence* carefully considers events and actors as they are enmeshed with sociopolitical and cultural histories to theorize how violence is made and unmade through and because of assemblages.

Although the entanglements that make up an assemblage of violence are difficult, if not impossible, to separate, this book focuses on two all-too-consistent recipients of violence: girls and women, and LGBTQ+ people. Attending to the case of women is important because, despite movements around the globe that aim to engender safety for women and girls, the United Nations still reports that an estimated one in three (35%) women worldwide have experienced physical and/or sexual violence at some point in their lives (UN Women, 2019). Although this percentage is already staggering, it is important to remember that women face other indelible forms of violence, ranging from emotional abuse to trafficking, from forced marriage to female genital cutting (World Health Organization, 2019). Further, most of the abuse that girls and women experience across the globe are forms of intimate partner violence, with almost one third of women who have been in a relationship reporting that they have experienced some form of sexual, emotional, or physical violence by their partner (World Health Organization, 2019).

In the case of LGBTQ+ communities, violence against queer-identifying and queer- identified people and groups continues to rise globally. As of 2020, some countries still enact the death penalty for same-sex relations (see, e.g., the "gay purge" that continues largely unabated in Chechnya), "corrective rapes" for lesbians abound (e.g., in El Salvador and South Africa), and thousands of trans people are maimed and killed annually across the globe (Advocates for youth, 2019; UN Human Rights Council, 2015). In the United States, several states continue to allow conversion therapy,[2] do not include violence against queer people under hate crime laws,[3] and do not have legal protections for LGBTQ+ individuals at work or in housing. While many countries have made strides towards equity and access for LGBTQ+ people, even within those countries people who have homo- and transphobias continue to create spaces where hatred and violence against queer bodies can bloom.

With these staggering statistics in mind, this book grounds dialogues about violence in the two interpretive studies mentioned in the Atithi Devo Bhava: one with women in India who were survivors of intimate partner and sexual violence,[4] and another with queer youth of color at a small middle school on the East Coast of the United States. The methodological differences between these two studies caused narratives to surface in different ways, consistent with each method's distinctive characteristics. During my time in India, I conducted a narrative inquiry (Clandinin, 2016; Clandinin & Connelly, 2000) that attended to questions of race (Delgado, 1989; Solórzano & Yosso, 2002) and critical race feminisms (Bhattacharya, 2017). After my initial visit to the shelter, I followed up with the director by phone and email several times. Although I have stayed in this area of the country several times because, spending months visiting people and becoming familiar with the culture, I was only able to spend relatively little time at the shelter itself. This created a context where participants may have been left feeling reluctant to speak vulnerably to a researcher whom they did not know well. While I worked hard to deepen my relationships, factors like language barriers and questions of safety for participants remained in the foreground of this work.

In contrast, the data at the middle school was collected through a sonic ethnography (for more on this method, see Gershon, 2017). Sitting at the intersection of sound studies and ethnography, sonic ethnographic is a tool that allows the researcher to listen deeply (Oliveros, 2005) to participants while attending theoretically, methodologically, and practically to how one can conceptualize a context while intentionally moving away from Western, ocular ways of knowing, being, and doing (e.g., Aoki, 1991; Erickson, 2004; Gershon, 2011). Students recorded their own narratives several times over the course of two years. This method allowed time for students to develop relationships with one another and with me in my roles as researcher and facilitator of their extracurricular club. The closeness that I developed with students could not be fostered through the narrative inquiry process used in India due to differences in time spent with participants. Finally, it is important to note that, regardless of the method, participants were brave in sharing these narratives and although this data has been deidentified, participants were asked multiple times for permission to share these sensitive stories.

Aside from methodological differences, there are significant contrasts between these two contexts and, as a result, also between how violence, aggression, and agency are enacted in each. Although this book discuses physical, verbal, and emotional violence against people in both places, other factors are important to consider. First, many of the participants in India identified as property of their husbands and other family members, especially their fathers. This feeling was discussed as a pervasive sociocultural norm, one with a historical legal basis that still remains in many parts of contemporary culture. Participants in the middle school, in contrast, articulated the ability to more easily enact their agency within their family structures. Second, in India all participants identified as cisgender and straight, while participants in the middle school on the East Coast had far more diverse identities in terms of gender and sexual orientation.

However, while different in expression and response, these seemingly disparate groups of people are similar in kind. Both studies show forms of deep violence and aggression against these marginalized groups, which in both cases tend to be permissible socioculturally within their contexts. This means that they are both also participants in cycles of violence that are impossible to enact without others' active sanctioning of violent behavior, including explicit looking away and implicit sanction through ongoing, knowing silences.

Each chapter of this text speaks about both theoretical and practical concerns. In the symphony that is our lives, theories exist as one way to articulate our experiences, whether they be quotidian or extraordinary. From Foucault's (1978) discussion of power to Bourdieu's (1997) scholarship on fields, our intra-actions within and across contexts are enmeshed with theories that, in turn, seek to explicate intra-actions. This does not mean that naming a theory or pairing events with theoretical ideas somehow lends legitimacy to one's experiences: the proverbial tree would still fall in the forest without a researcher naming its descent. However, sometimes in attending to and naming an event or context, an experience can be heard differently. For example, racism has always impacted the bodies of the oppressed but by naming racial battle fatigue (Smith et al., 2007), dialogues around normalized hostilities have become increasingly recognized across communities—from psychologists to the academy (Smith, 2014). This text similarly positions the theoretical framework of assemblages of violence as knotted within research contexts and participants' narratives.

Scientific Entanglements: The Mattering of Voices

The attraction and repulsion of bodies as they shape phenomena is central to all meaning-making processes. This is not solely an attention to the fact that bodies *are*, but rather a focus on how and what bodies *become* (Bennett, 2010; Puar, 2013) and on what they produce through relationships, for no kind of body, event, or trajectory is ever *not* in relation. Experiences are always enmeshed with local and less-local bodies that coproduce an endless multiplicity of ontoepistemologies (Bennett, 2007; Gilmore, 2002; Haraway, 2008; Snaza et al., 2016). Attending to how ways of being, knowing, and doing are made and unmade through entangled bodies is not new: from Anna Julia Cooper's (1892) discussion of the intersections of race and gender, to Shirley Graham DuBois' (1932) musical drama, *Tom-Tom*, which follows the complex rhythms that ebb and flow through history across narratives, spaces, and places. Further, such dialogues span disciplines, from Barad's (2007) new materialist scholarship on the coconstitutive nature of relations—one that argues the significance of how intra-actions are responsible for the production of agency—to Massey's (2005) work in critical geography focused on the multiplicity of space—where she explicates that "space" is constituted through a multiplicity of unfolding relations that are always in the process of being produced between bodies.

Given that knots of bodies and events are so significant to meaning making or, as others have argued, to the "mattering" of things (e.g., Barad, 2007; Tsing, 2005; Woodson, 1933), there are several trajectories for such dialogues. Anthropologists like Clifford Geertz (1973) have imagined the "webs of significance" that impact people, places, and events, while others like Dorinne Kondo (1990) have discussed the complexities of "violent attempts to capture meaning" (p. 7) through complex contexts. Carter G. Woodson (1933) explicated the connections between institutions as metaphors that become the realities that affect bodies, while Christina Sharpe (2016) focuses on the contemporary waters that are haunted and shaped by histories. How bodies are shaped, and how they affect and are affected by the everyday, by the local and far away, by the predicable and the events one can never begin to imagine, matters and is always already matter*ing*; this is the constant ongoing becoming of things and how they are perceived.

One way to describe the knotted relationships that are entwined with meaning-making process as bodies become is through the language and concepts of quantum mechanics. Although the vast field of quantum theory has yielded several salient dialogues on this topic as it relates to social sciences (e.g., Barad, 2007; Kuhn, 1987; Rautio, 2014; Rosiek, 2018), here I focus on one facet of the field: quantum entanglements. While there are several inroads to discussing the entangledness of the everyday, I use quantum theory for (at least) the following reasons. First, as Barad (2007) argues, "matter and meaning are not separate elements. They are inextricably fused together, and no event, no matter how energetic, can tear them asunder" (p. 3). To Barad's point, affects are grounded in the material. Similar to Ahmed (2010), Berlant (2011), Stewart (2007), and other scholars' conceptualizations, affect here is understood as more than just a set of emotions that "stick" (Ahmed, 2010) on, between, and through bodies. Drawing on these insights throughout this book, I argue that affect is a sociopolitical tool that is often used to reify power. This is why, for example, racial battle fatigue has physical consequences on the body, and why ordinary affects (Stewart, 2007) of normalized biases like racism can set events into motion.

Second, research is always political. As such, using quantum theory as a metaphor to better understanding the intra-actions across layers of empirical data becomes a political act of reclaiming the positionality of social science within and against other forms of scientific research (Cannella & Lincoln, 2004; Williams, 2012). In other words, this is a rejection of the "hard" and "soft" framing of the sciences that still impacts institutions and preserves the cis-hetero patriarchal connections that pervade educational contexts (Hedges, 1987; Morgans & Ryan, 2019). From a queer feminist perspective, it is equally significant for the narratives included in this text—those of women and queer youth—to be a queered space (Halberstam, 2005; Wozolek & Mitchell, 2018). In other words, I intend for this text to be a place where several scientific fields are used to think deeply about the multiple bodies, both human and nonhuman, that are central to the lived experiences of the participants.

Finally, this is not an attempt to bolster one field or deconstruct another. Rather, in conversation with other curriculum theorists who regularly theorize through interdisciplinary approaches, I continue the work of engaging across disciplines. Weaving disciplines together is certainly not new and there are several well-trodden dialogues regarding the mutual ideas that cross and connect fields (e.g., Agar, 2004; Barad, 2007; Trnka & Lorencova, 2016; Gershon, 2008). This work also attends to the fact that sciences, whether the work of an anthropologist or a physician, have histories of physical and psychological harm to research participants. Although these offenses are not equal in their violence (Mengele committed far more harmful atrocities than early ethnographers studying brown and Black participants, for example), it is important to attend to the degrees of violence that exist across fields. My sewing together of the sciences is therefore both a political act of reclaiming space and a way of paying attention to violence in and across disciplines. I do so with attention to the constant and continued physical, sociocultural, and emotional harm that the "hard" sciences have inflicted globally. This book therefore moves across fields to think about the connections between bodies as they move, bond, divide, act, and react as they become through entanglements.

Threads of a Knot

Much like Nespor's (1997) description of schooling, where he argues for the importance of attending to "flows rather than states…and the layered connections that knot [places, spaces, and events] together rather than on simple linear histories of circumscribed events or settings" (p. xiv), this book attends to the messy, complex nature of experiences through the two studies introduced above. The purpose of this book is to theorize how violence is created, moves, and is maintained through an assemblage. At the same time, the book aims to show how factors like hope, love, and joy are central to experience and, in many ways, also exist within violent intra-actions.

Chapter 1 focuses on hope as it frequently manifests within an assemblage through quantum entanglements. Specifically, I explore the local and less local correlations between bodies, in this case bodies of hope. In the chapter, I attend to the ways that quantum mechanics offers a helpful framework for describing intra-actions and affects, both as a science that describes the spooky, wacky, and often hard-to-name behaviors of the universe and as a way of theorizing the big through the very small. However, while this book draws heavily from quantum physics to think about the theoretical and material groundings of intra-actions, it is also deeply entangled with social theory in its methodological approach and analysis.

Additionally, in this chapter I describe the connections between quantum physics, social theory, and anthropology with the utmost care for several reasons. First, and most important, is a question of ethics and (un) intentional harm. I emphasize the historical underpinnings of oppression within the fields of science and anthropology, which have been physically and emotionally harmful to people and groups. Second, I attend to the way these fields overlap and diverge: On one hand, while there are clear theoretical similarities, quantum physics is not synonymous with social theory. On the other, the world is a complex place, especially when one gets down to the interrelations of human beings. Quantum physics and social science both explore such complexities through their respective lenses, and I use both in this text to explore the interrelations of contexts and events.

The construct of "fields", for example, is a theoretical concept that is central in both social science and quantum physics (e.g., Berkson, 2014; Bourdieu, 1996, 1997; Burawoy, 2014; Hartmann, 2001). Where quantum field theory discusses how particles act, react, and interact across systems with an infinite number of degrees of freedom (Stanford Encyclopedia of Philosophy, 2012), sociology mobilizes fields to examine how individuals and their social positions are located (Bourdieu, 1996). In this book, I speak across both quantum and social theories to understand the power dynamics in fields as they are related to questions of agency, fluidity, and boundaries of intra-actions.

Quantum entanglements exist within and across and assemblages. Zooming out from the entanglements, Chapter 2 documents the contours of an assemblage of violence. The image for this chapter is a Gordian knot, something that is used to describe a complex or unsolvable problem. Although it is impossible to fully tease apart the knot that is an assemblage, this chapter maps the contours of violent assemblages in order to think about how normalized aggression can simultaneously move, bind, and liberate people and communities. Similar to Massey's (2005) discussion of the multiplicity of space, this chapter attends to the multiplicity of violence that undergirds spaces. Thinking through these complex interrelations of violence is especially important in research with survivors of intimate partner violence and queer youth of color. Such assemblages, to borrow from Weheliye (2014), are central to the oppressive realities of those marginalized by violence.

As one can imagine, tracing trajectories of violence is difficult. Chapter 3 introduces pathology as a method to begin to trace the movements of an assemblage. This is an intentional move toward the well-trodden language of Deleuze and Guattari's (1987) dialogue on rhizomes and assemblages. To date, scholarship has tended to focus on what a rhizome is, or occasionally on what it does, but there is little discussion of how rhizomes move beyond noting their complexity and seemingly random nature. Pathologies are helpful because they trace such fluid movements without reducing their complexity. This chapter elaborates how pathology might be used methodologically to trace the affects that flow through the lives of marginalized and oppressed people and groups.

Finally, Chapter 4 argues that assemblages of violence form and inform a curriculum that exists both inside and outside formal educational contexts. Using the rich histories present in curriculum studies, I argue that violence is just as present in what students are intended to learn as it is in those lessons that are (un)intentionally absent from the curriculum. Violence is learned across spaces and places in the school, from the classroom to the schoolyard, and is nested in almost every interaction one has in the school and throughout systems of schooling. As Woodson (1933) argues, violence too often starts in the schoolroom. Extending this insight further, I trace how violence leaks from the schoolroom into communities and how it is maintained by pedagogies of violence. Chapter four joins long-standing dialogues on the multiplicity of violence as it is taught and learned across educational contexts in ways that sticks on bodies and beings (e.g., Cooper, 1892; Freire, 1970; Gilbert, 2014; Miller, 2005; Morris, 2016; Woodson, 1933). Building on these voices, I explain how violent assemblages not only normalize violence but also teach how one moves within, against, and across violence.

Returning to my opening narrative, "Unsnarling Chains", this book attends to the entanglements found in knotted ontoepistemologies (Barad, 2007) in assemblages of violence. I do not offer solutions to such assemblages because, as any person attending to systemic inequities and oppressions knows, answers and silver bullet approaches generally harm more than they help. Rather, my intention in this book is to be vulnerable with the entanglements, to witness their impacts, and to begin to unfold their implications. If you don't mind feeling excited yet challenged in ways that lead to a Muppet-scream-dance cycle, I invite you to join me in a journey into the symphony of violent assemblages.

Notes

1 Dialogues on space and place have become rather common in social sciences, including education (e.g., Helfenbein, Ayers, & Buendía, 2017), critical geography (e.g., Massey, 2005), and other qualitative research (e.g., Tsing, 2005). My dialogue on space and place argues for the significance of multiplicity within space and the sociopolitical and cultural norms and values imbricated within space and place. I return to this discussion in Chapter 2.

2 See https://www.lgbtmap.org/equality-maps/conversion for the latest information on states that do not ban conversion therapy. As of 2020, 29 states and four U.S. territories did not have state laws or policies against conversion therapy.

3 See https://www.hrc.org/state-maps/hate-crimes for the latest information on states that do not include LGBTQ+ individuals in their hate crime statutes. As of 2020, 18 states did not protect LGBTQ+ citizens under hate crime laws and only 20 states included gender identity along with sexual orientation under their laws.

4 This text uses the term "intimate partner violence" to emphasize that violence can occur between people, regardless of sexual orientation, gender expression or identity, social class, or country of origin, or other factors in contexts of violence. For victims and survivors of intimate partner violence, know that you are not alone and that help can be found at https://www.thehotline.org/ or by calling 1.800.799.7233 in the United States, or at https://www.domesticshelters.org/resources/national-global-organizations for those seeking help internationally.

References

Advocates for Youth (2019). Lesbian, gay, bisexual, and transgender (LGBT) youth in the global south. Retrieved from: https://advocatesforyouth.org/resources/fact-sheets/lesbian-gay-bisexual-and-transgender-lgbt-youth-in-the-global-south/

Ahmed, S. (2010). Happy objects. In M. Gregg & G. J. Seigworth (Eds.), *The affect theory reader* (pp. 29–51). Durham, NC: Duke University Press.

All Things Considered. (2019). Cellist Yo-Yo Ma plays Bach in shadow of border crossing. https://www.npr.org/2019/04/13/713092703/cellist-yo-yo-ma-plays-bach-in-shadow-of-border-crossing

Altenmüller, E. O. (2004). Music in your head. *Scientific American Mind, 14*(1), 24–31.

Aoki, T. T. (1991). Sonare and vidare: Questioning the primacy of the eye in curriculum talk. In G. Willis, & W. H. Schubert (Eds.), *Reflections from the heart of educational inquiry: Understanding curriculum and teaching through the arts* (pp. 182–198). Albany, NY: State University of New York Press.

Barad, K. (2007). *Meeting the universe halfway: Quantum physics and the entanglement of matter and meaning*. Durham, NC: Duke University Press.

Bakhtin, M. M. (1981). *The dialogic imagination: Four essays*. (Translated by C. Emerson & M. Holquist). Austin, TX: University of Texas Press.

Bennett, J. (2010). Vibrant matter: A political ecology of things. Duke University Press.

Berkson, W. (2014). *Fields of force: The development of a world View from faraday to einstein*. New York: Routledge.

Berlant, L. (2011). *Cruel Optimism.* Bedford, O., & Hwang Durham, NC: Duke University Press.

Bettmann, O. L. (1986). MUSIC: Bach the rhetorician. *The American Scholar*, 55(1), 113–118.

Bhattacharya, K. (2017). *Fundamentals of qualitative research: A practical guide*. New York: Routledge.
Boon, J. P., Noullez, A., & Mommen, C. (1990). Complex dynamics and musical structure. *Journal of New Music Research*, *19*(1), 3–14.
Bourdieu, P. (1996). *The rules of art: Genesis and structure of the literary field*. Standford, CA: Stanford University Press.
Bourdieu, P. (1997). *Outline of a theory of practice*. Stanford, CA: Stanford University Press.
Burawoy, M. (2014). Introduction: Sociology as a combat sport. *Current Sociology*, *62*(2), 140–155.
Bylsma, A. (1992). *J.S. Bach suites for violoncello solo, BVW 1007-1-12*. New York: Sony.
Cannella, G. S., & Lincoln, Y. S. (2004). Epilogue: Claiming a critical public social science—Reconceptualizing and redeploying research. Qualitative Inquiry, 10(2), 298–309.
Clandinin, D. J. (2016). *Engaging in narrative inquiry*. New York: Routledge.
Clandinin, D. J., & Connelly, F. M. (2000). *Narrative inquiry: Experience and story in qualitative research*. San Francisco, CA: Jossey-Bass Publishers.
Cooper, A. J. (1892). *A voice from the south (by a black woman from the south)*. Xenia: OH: Aldine Printing House.
Deleuze, G., & Guattari, F. (1987). *A thousand plateaus: Capitalism and schizophrenia*. Minneapolis, MN: University of Minnesota Press.
Delgado, R. (1989). Storytelling for oppositionists and others: A plea for narrative. *Michigan Law Review*, *87*(8), 2411–2441.
Erickson, F. (2004). *Talk and social theory: Ecologies of speaking and listening in everyday life*. Cambridge, UK: Polity Press.
Feld, S. (1995). From schizophonia to schismogenesis: The discourses and practices of world music and world beat. In G. E. Marcus, & F. R. Myers (Eds.), *The traffic in culture: Refiguring art and anthropology* (pp. 96–126). Berkeley, CA: University of California Press.
Foucault, M. (1978). *The history of sexuality*. New York: Pantheon Books.
Freire, P. (1970). *The pedagogy of the oppressed*. New York, NY: Continuum Books.
Fuchsman, K. (2019). Gaslighting. *The Journal of Psychohistory*, *47*(1), 74–78.
Futrell, R., Simi, P., & Gottschalk, S. (2006). Understanding music in movements: The white power music scene. *The Sociological Quarterly*, *47*(2), 275–304.
Gershon, W. S. (2011). Introduction: Towards a sensual curriculum. *Journal of Curriculum Theorizing*, *27*(2), 1–16.
Gershon, W. S. (2017). *Sound curriculum: Sonic studies in educational theory, method, & practice*. New York, NY: Routledge.
Gershon, W. S. (2018). Reverberations and reverb: Sound possibilities for narrative, creativity, and critique. *Qualitative Inquiry*, 1–11.
Geertz, C. (1973). *The interpretations of cultures*. New York: Basic Books.
Gilbert, J. (2014). *Sexuality in school: The limits of education*. Minneapolis, MN: University of Minnesota Press.
Goodman, S. (2012). *Sonic warfare: Sound, affect, and the ecology of fear*. Cambridge, MA: MIT Press.
Gould, C. S., & Keaton, K. (2000). The essential role of improvisation in musical performance. *The Journal of Aesthetics and Art Criticism*, *58*(2), 143–148.

Gregg, M., & Seigworth, G. J. (Eds.) (2010). *The affect theory reader*. Durham, NC: Duke University Press.

Halberstam, J. (2005). *In a queer time and place: Transgender bodies, subcultural lives*. New York: New York University Press.

Haraway, D. (2008). *When species meet*. Minneapolis, MN: University of Minneapolis Press.

Harris, K. L. (2018). Mapping gender and violence: Describing reality, resisting abuse. *Women's Studies in Communication*, *41*(2), 113–116.

Hartman, S. V. (1997). *Scenes of subjection: Terror, slavery, and self-making in nineteenth-century America*. New York, NY: Oxford University Press.

Hartmann, S. (2001). Effective field theories, reductionism, and explanation . *Studies in History and Philosophy of Modern Physics*, *32*(2), 267–304.

Hedges, L. V. (1987). How hard is hard science, how soft is soft science? The empirical cumulativeness of research. American Psychologist, 42(5), 443.

Helfenbein, R., Ares, N., & Buendía, E.. (2017). *Deterritorializing/reterritorializing: Breakthroughs in the sociology of education*. Boston, MA: Sense.

Kondo, D. (1990). *Crafting selves:* Power, gender and discourses of identità in a Japanese workplace. Chicago, IL: University of Chicago Press.

Kuhn, T. S. (1987). Black-body theory and the quantum discontinuity, 1894–1912. Chicago, IL: University of Chicago Press.

LaBelle, B. (2018). *Sonic agency: Sound and emergent forms of resistance*. London, UK: Goldsmiths Press.

Levi, E. (1990). The Aryanization of music in Nazi Germany. *The Musical Times*, *131*(1763), 19–23.

Love, B. L. (2016). Complex personhood of hip hop & the sensibilities of the culture that fosters knowledge of self & self-determination. . *Equity & Excellence in Education*, *49*(4), 414–427.

Massey, D. (2005). *For space*. Thousand Oaks, CA: Sage Publishing.

Massumi, B. (2002). *Parables for the virtual: Movement, affect, sensation*. Durham, NC: Duke University Press.

McCauley, H. L., Bonomi, A. E., Maas, M. K., Bogen, K. W., & O'Malley, T. L. (2018). #MaybeHeDoesntHitYou: Social media underscore the realities of intimate partner violence. *Journal of Women's Health*, *27*(7), 885–891.

Miller, J. (2005). *The sounds of silence breaking: Women, autobiography, curriculum*. New York, NY: Peter Lang Publishing, Inc.

Morgans, A. K., & Ryan, C. J. (2019). Hard Problems Need" Soft" Science: Integrating Quality of Life into Treatment Decision Making. European urology, 75(6), 948.

Morris, M. W. (2016). *Pushout: The criminalization of black girls in schools*. New York, NY: The New Press.

Moten, F. (2003). *In the break: The aesthetics of the black radical tradition*. Minneapolis, MN: University of Minnesota Press.

Nespor, J. (1997). *Tangled up in school: Politics, space, bodies and signs in the educational process*. Mahwah, NJ: Lawrence Erlbaum Associates Inc.

Oliveros, P. (2005). *Deep listening: A composer's sound practice*. Lincoln, NE: Deep Listening Publications.

Peretz, I., & Coltheart, M. (2003). Modularity of music processing. *Nature neuroscience*, *6*(7), 688–691.

Pink, S. (2007). *Doing visual ethnography* (2nd ed). London, England: Sage.

Puar, J. (2013). 'I would rather be a cyborg than a goddess': Intersectionality, assemblage, and affective politics. *Meritum, 8(2),* 371–390.

Rautio, P. (2014). Mingling and imitating in producing spaces for knowing and being: Insights from a Finnish study of child–matter intra-action. Childhood, 21(4), 461–474.

Riggs, D. W., & Bartholomaeus, C. (2018). Gaslighting in the context of clinical interactions with parents of transgender children. *Sexual and Relationship Therapy, 33*(4), 382–394.

Rosaldo, R. (1993). *Culture & truth: The remaking of social analysis*. Boston, MA: Beacon Press.

Rosiek, J. L. (2018). Agential Realism and Educational Ethnography. In D. Beach, C. Bagley, & S. Marques da Silva (Eds.), The Wiley handbook of ethnography of education (pp. 403–423). Hoboken, NJ: John Wiley & Sons, Inc.

Sharpe, C. (2016). *In the wake: On blackness and being.* Durham, NC: Duke University Press.

Siblin, E. (2010). *The Cello Suites: JS Bach, Pablo Casals, and the Search for a Baroque Masterpiece*. Toronto, ON: House of Anansi.

Smith, W. A. (2014). *Racial battle fatigue in higher education: Exposing the myth of post-racial America*. Landham, MD: Rowman & Littlefield.

Smith, W. A., Allen, W. R., & Danley, L. L. (2007). "Assume the position… you fit the description" psychosocial experiences and racial battle fatigue among African American male college students. *American Behavioral Scientist, 51*(4), 551–578.

Snaza, N., Sonu, D., Truman, S. E., & Zaliwska, Z. (Eds.). (2016). Pedagogical matters: New materialisms and curriculum studies. New York: Peter Lang Publishing, Inc.

Solórzano, D. G., & Yosso, T. J. (2002). Critical race methodology: Counter–storytelling as an analytical framework for education research. *Qualitative inquiry, 8*(1), 23–44.

Spitta, P. (1951). *Johann Sebastian Bach: his work and influence on the music of Germany 1685– 1750*. Translated by C. Bell and J. A Fuller-Maitland. New York: Dover Publications. Stanford Encyclopedia of Philosophy. (2012). Quantum Field Theory. Retrieved from: https://plato.stanford.edu/entries/quantum-field-theory/#Intro

Sterne, J. (2012). *The sound studies reader*. New York, NY: Routledge.

Stewart, K. (2007). *Ordinary affects*. Durham, NC: Duke University Press.

Stoller, P. (1997). *Sensuous scholarship*. Philadelphia, PA: University of Pennsylvania Press.

Sweet, P. L. (2019). The sociology of gaslighting. *American Sociological Review, 84*(5), 851–875.

Tobias, H., & Joseph, A. (2018). Sustaining systemic racism through psychological gaslighting: Denials of racial profiling and justifications of carding by police utilizing local News media. *Race and Justice*, 1–32.

Tsing, A. L. (2005). *Friction: An ethnography on global connection*. Princeton, NJ: Princeton University Press.

UN Women (2019). Facts and figures: Ending violence against women. Retrieved from: https://www.unwomen.org/en/what-we-do/ending-violence-against-women/facts-and-figures

UN Human Rights Council (2015). Report of the United Nations High Commissioner for Human Rights on Discrimination and violence against individuals based on their sexual orientation and gender identity. Retrieve from: https://www.ohchr.org/EN/Issues/Discrimination/Pages/LGBTUNReports.aspx

Waltham-Smith, N. (2006). Disruptive spatiality and the experience of recordings of bach solo cello suites. *Current Musicology*, *82*(2), 33–59.

Weheliye, A. G. (2014). *Habeas Viscus: Racializing Assemblages, Biopolitics, and Black Feminist Theories of the Human*. Durham, NC: Duke University Press.

Williams, M. (2012). Science and social science: An introduction. New York: Routledge.

Winberg, M.. (2019). Can you hear it? Sonic devices play high-pitched noises to repel teens, *National Public Radio*. https://www.npr.org/2019/07/10/739908153/can-you-hear-it-sonic-devices-play-high-pitched-noises-to-repel-teens

Winold, A. (2007). *Bach's cello suites, volumes 1 and 2: Analyses and explorations*. Indianapolis, IN: Indiana University Press.

Woodson, C. G. (1933). *The Mis-Education of the Negro*. Washington, DC: The Associated Publishers.

World Health Organization. (2019). Violence against women: Key facts. Retrieved from: https://www.who.int/news-room/fact-sheets/detail/violence-against-women

Wozolek, B. (2018). Gaslighting queerness: Schooling as a place of violent assemblages. *Journal of LGBT Youth*, *15*(4), 319–338.

Wozolek, B. & Mitchell, R.P. (2018). Plastic or Phalloplasty: Negotiating Masculinity and (Cis)gender norms in schools and the academy, *The Journal of Curriculum Pedagogy,* 15(3), 318–322.

1

WHEN THINGS ATTRACT...OR DON'T: INTRA-RELATIONS AND QUANTUM ENTANGLEMENTS

> *"You know what the [talent] show did for me? It made me feel good. Like real good. Not just about school but about me. People clapped. For me. For us. Did you see that? I gotta admit, maybe there is hope for this place. For us. If people could see us that way then maybe they can see us that way all the time..."*
>
> —*Dominique, Grade 7*

We lead entangled lives. One only needs to reflect on a moment when they felt at one with another body, when art evoked extreme and perhaps unexpected emotion, or how the sensation of sun-kissed skin lingers long after sunset. It has been well documented that twins regularly experience such entanglements through an awareness of each other's affects, most notably through sensing pain and crisis across great distances (Anderson, 2019; Kubis & Kouke, 1937; Parker & Jensen, 2013). The very fibers of twins' "being" seem impossibly entwined in ways that psychologists and medical professionals still have yet to fully grasp.

While those who study this phenomenon tend to focus on the narratives of twins, people who have experienced deep emotions for others, whether positive or negative, have also reported knotted ways of being that are inextricable from their counterpart. One example is Derrida's (1988) enigmatic essay in which he seeks to disrupt the notion that identity is stable in terms of an "I" that is separate from the "other". While anthropologists and sociologists have well-trodden dialogues on how one's identity is layered with how they are identified in and through local and less local communities (e.g., Coffey, 1999; Goffman, 1963; Rahder, 2020; Varenne & McDermott, 1998), Derrida was arguing for a much more entangled understanding of how one "is" in the world, a way of being that is enmeshed with others.

This chapter addresses how everyday affects are not only knotted between nearby bodies, both human and nonhuman, but also are sometimes entangled so deeply that these affects effectively communicate across distances, or through quantum entanglements. Throughout, I refer to "bodies" quite a bit to express their movements, agencies, and coconstituted subjectivities. For clarity, it is important to tease apart the many bodies being discussed. Similar to Hultman and Lenz-Taguchi (2010), I employ the notion that bodies are "forces that overlap and relate to each other" (p. 529). This relational materialism will be further explained below using Barad's (2007) discussion of intra-action. What is important at this point is that humans do not exist in a silo; they are constantly in intra-action with things like other animals,[1] objects, affects, and cultures. The human body is therefore as significant as the bodies of cultures with which it regularly intra-acts.

By culture, I mean the ideas and ideals that are normalized and valued within and between groups of people, which manifest in a multiplicity of expressions (Page, 2000; Visweswaran, 2010). By using "body of culture", however, I turn to a relational understanding of bodies, one in which culture is agential (Hultman & Lenz-Taguchi, 2010). As I argue throughout this chapter, this means that culture has the ability to act in ways that form and inform other bodies that are in relation to it. While the bodies themselves are significant, I primarily focus on their relations and relationships that render them agential. Following Puar (2013), my primary focus is not just on what bodies *are*, but on what they *do*.

While the next chapter focuses on the assemblage as a whole, this chapter theorizes what happens within an assemblage, attending specifically to the intra-actions that are central to an assemblage's entanglements. I begin by explaining several facets of quantum entanglements, starting with how bodies become entangled. I then use this theory to think about New Horizons, specifically discussing the details of the school's talent show. It was at that show, as the epigraph above hints that I observed how hope can move and spread in and between groups of people. Following this discussion, I share the narrative of Prisha, one of the women I met in India. Her narrative exemplifies questions of entanglement, showing how people become enmeshed and moved by broader cultural norms and values. Together, these stories help us better understand how affects like hope or despair move—theorization that is deeply necessary. While on the surface this might seem straightforward, ask yourself: how does an affect like hope move? Tacking back and forth between theory and participant narratives,

I produce a complex understanding of how affects move, and how human animals and ecologies affect and are affected by one another. The complexity of such movements is explored in the following section.

Agentic Entanglements

Generally defined as a state of being wrapped or twisted in a relational mass—often in ways that make it difficult to extricate one part of the knot—entanglements are one way to understand the relationships and onto epistemologies born from experiences. Discussions of entanglement are often based in the complexity of relations, frequently turning to the study of causes and effects within the intricate imbrications of bodies. Across disciplines, the study of entanglements and the "causes" for such effects has become common. For example, affect theorists have discussed the "forces of encounter that transpire within and across the subtlest of intensities… [that become] a palimpsest of force-encounters traversing the ebbs and swells of intensities that pass between 'bodies'" (Gregg & Seigworth, 2010, p. 2). Qualitative researchers have similarly thought about the complex relationships between bodies that are central to meaning-making processes (e.g., Tsing, 2005), and educational theorists have attended to the knot of relations that is central to schooling (e.g., Nespor, 1997; Varenne & McDermott, 1998). In short, what it means to be and to become through entanglement continues to incite imagination across fields of research.

One way to theorize entanglements and, by extension, assemblages, is through Karen Barad's (2007) framing of intra-action. Barad distinguishes between *inter-* and *intra-*action, noting that during *inter*actions, both human and nonhuman bodies maintain a level of independence. Each entity exists and, in many ways, will continue to exit with separate identities and agencies that remain intact before and after an interaction. In *intra*-action, individuals materialize through encounters. This is not to say that bodies do not have agency or ways of being and knowing prior to intra-action but, rather, that bodies engage in coconstitutive ways that allow agency and a new sense of being to emerge from the relationship, not from outside of it.

Take, for example, the Stonewall uprising of 1969. One can say that Stonewall was not just an event but an intra-action of human and nonhuman parts. This includes human bodies, discourses on queerness, homo- and transphobia, politics, news coverage on civil rights, and other such factors. Stonewall was therefore a phenomenon made and unmade through intra-actions between the event, culture, and the dissemination of information about the event. Through intra-actions, people were brought together through the phenomenon of Stonewall, yet these intra-actions also separated actors into new coconstituted subjects: people became, at least temporarily, those rioting and those choosing not to join the demonstrations, those impacted through physical proximity and those not affected, and those exposed or not exposed to queerness, to name a few.

In this way, everyone in the country could be considered entangled in Stonewall, although, certainly, not everyone directly interacted with the riots in any fashion they would recognize. While I placed the above categories of being into false binaries to illustrate the point, there are infinite points between each of these positions and a multiplicity of ways that these positions can collapse on themselves. It is quite possible that a queer body might not have been in physical proximity to Stonewall but still intra-acted with the national media coverage. Although an individual might choose to remain absent from and silent within conversations surrounding events like the riots, intra-action occurs regardless of awareness or participation.

Even before it was a national touchstone, whether people were aware or not of Stonewall and its place in understanding queer identities and rights, these understandings were nonetheless part of local and less-local constructions about what it might mean to be human in relation to national politics and intimate partner relationships. Or, in Barad's terms, regardless of whether one interacted with or was aware of intra-actions with the event, the ecology of the United States intra-acted with Stonewall, therefore distributing responsibility for the matter produced within these intra-actions—namely discourses, materials, and subject positions. This is not unlike the intra-actions of the #MeToo movement, the recent COVID-19 pandemic, or the way that hopes emerged from the New Horizons talent show. Regardless of the event, through intra-actions, agency and ways of being, knowing, and doing are emergent.

This is where agency comes into play. To date, there have been many ways to think about agency (e.g., Bourdieu, 1973; Foucault, 1978; Goffman, 1961; Ortner, 2006). The long history of discussing agency across fields has created many important contributions to our understanding of how bodies move and act within and against systems of power. Here, I focus on two particular conceptualizations of agency through the respective scholarship of Sherry Ortner (2006) and Karen Barad (2007). While both Ortner and Barad argue for an agency that is produced through relationships, they diverge in their arguments about how that agency is constructed.

As an anthropologist, Ortner frames her discussion of agency within human interactions and with attention to intentionality and power. She begins with intentionality, a concept that she defines as being on a continuum of levels of consciousness that are "directed toward some end that might include highly conscious plots and plans [to] somewhat more nebulous goals [to] desires, wants, and needs" (Ortner, 2006, p. 134). Ortner explicates that some conversations about agency miss intentionality while others make intentionality too central. She instead argues that, while intention is important to agency, there is a need to recognize the "strong role of active (though not necessarily fully 'conscious') intentionality in agency that differentiates agency from routine practices" (Ortner, 2006, p. 138). In other words, there is a difference between what one does daily and what one does with a sense of intention, however unconscious that intention may be. Or, conversely, one can enact their agency regardless of their awareness about their own intentions toward particular ends.

Finally, Ortner argues that agency exists is in relation to power, which, in turn, can be understood as strongly controlling but never complete. While enacting agency is an act of power, Ortner writes, actors still live within politically and culturally controlled local contexts. Take for example an intimate partner violence counseling center like the one in India, which exists in a context where reporting abuse is not a cultural norm. The center might have been organized with a sense of purpose for that particular place, with clear attention to the issues that survivors of abuse might face in the community and an intention to disrupt cycles of violence. Those who work in the center are operating to disrupt violence on both local and less-local levels. However, the ability for counselors and survivors to enact agency is predicated not only on the immediate context but also on broader norms and values that are implicit and explicit across communities. Such norms might dampen the center's ability to be helpful, as well as survivors' ability to participate in services at the shelter.

The relationship between power and agency exists within what I call *agentic contingencies,* or the circumstances that impact a person's ability to use her agency. Agentic contingencies are fluid and, like Ortner's power or Bourdieu's (1973) fields, controlling but not all-encompassing. While the center for intimate partner violence in the example above may be concerned with survivors, counselors may be situated within agentic contingencies that include time, their power within and around the community, and their own relationships to the context. Counselors may additionally be limited or supported by factors like state and federal funding. An attention to agentic contingencies is important to understanding how some practices are normalized and how one might be able to resist or refuse (McGranahan, 2016; Ortner, 2006; Simpson, 2014) cultural values that allow violence to propagate.

Similar to McGranahan (2016) and Simpson (2014), I acknowledge that although resistance and refusal are linked, they are not synonymous. As a term, resistance tends to be fraught with binaries, such as domination versus resistance (Ortner, 2006) or hidden and indirect versus public engagement with power (Scott, 1990). Refusal can similarly fall into these binaries, but while resistance can be understood as an opposition to power, refusal is an explicit rejection of one political project for another (Graeber, 2013).

Refusal, McGranahan (2016) writes, "marks the point of a limit having been reached: *we refuse to continue on this way*" (p. 320).

As I have written elsewhere (Wozolek, 2020), agentic contingencies often underscore how resistance and/or refusal are tied to questions of power. A woman working at a center might have fewer agentic contingencies than her clients. The ability to refuse a patriarchal system of violence in ways that are "generative and strategic, mov[ing] toward one thing, belief, practice or community and away from another" (McGranahan, 2016, p. 319) might be far more possible for a center director than it is for a woman still living with/in a violent context. The client can, perhaps, actively resist systems of oppression, but she might not have the power to completely reject them in her life, a power that might be afforded to her abuser who is able to refuse her humanity.

Barad (2007) similarly discusses issues of power and agency as she uses intra-actions to think about how agency is produced through entanglements. Unlike Ortner, Barad emphasizes that the interdependence of bodies in a phenomenon produces agency while arguing that nonhuman bodies also have agency. In his work on racism, Jerry Rosiek (2018) reminds us that this conceptualization of agency "is not a claim that things have consciousness or make plans. It is, instead, a claim that reality of phenomena lies in the relations established through a process of intra-action" (p. 81). Rosiek attends to the ways that racism has agency in its movements and ability to shape sociocultural norms and phenomena. Returning to the example of Stonewall, Rosiek might argue that homo- and transphobias have agency and therefore moved through the context, shaping the phenomena created within and by the uprising.

At the intersection of Ortner's and Barad's scholarship is a sense that agency encompasses the intention and attention of bodies—human and nonhuman alike—as they intra-act within and against lines of power. Braiding these two conceptualizations together is important for at least two reasons. First, affects like racism do have agency, but that agency still exists within broader systems of power and their related agentic contingencies, which are carried out with a sense of intention. The same can be said for antiracist agency, which can move against normalized bodies of bias despite deep-seated histories of racism. Second, within intra-actions, agency is coconstituted but also, to borrow from Ortner, heavily dependent on the cultural ideas and ideals that exist with/in a given intra-action. This agential construction, as Ortner (2006) argues, is often a cultural construction that is "nourished or stunted" (p. 137) through lines of power. The imbrication of Ortner's and Barad's careful attentions toward agency, its coconstitutive nature, and the agentic contingencies that govern it are important to thinking about entanglements as they unfold throughout this book.

Entangling Quantum Mechanics

Up to this point, I have discussed how bodies function when they are already entangled. Here, I discuss some of the many ways and things with which bodies become entangled. Much in the way an improvising jazz musician plays across a bar line to create a rhythmic illusion in solo lines that don't resolve in more usual or anticipated places, the narratives here are enmeshed in certain ways to illustrate how bodies become entangled. To make this entwining more explicit, the solo lines of each story are marked in the text as they change, but they are not always resolved immediately.

As argued above, one way to think about entanglements is as the knotted layers through which bodies become so enmeshed that it becomes difficult, if not impossible, to extricate their original states. Following Barad's (2007) argument, when bodies intra-act they also entangle, impacting and shifting ways of being across and as a result of phenomena. The process of being made and unmade through an event is always unique. Even if the same bodies are involved, they would have already been changed from the first encounter, making the shift different for subsequent events. Further, where some entanglements take years, others are spontaneous. For example, while love can sometimes feel spontaneous, there are often also long-term relationships at play that were fostered over time and that helped form the two people who suddenly experience sensations of "love". Following entanglements, I am as interested in the instance when such feelings materialize as in the previous relationships that undergird that particular moment. In short, while all events involve entanglements, and both entanglements and events are present, each iteration of an entanglement is necessarily discreet unto itself.

This section explicates quantum entanglements for human intra-actions. Human relationships are unpredictable, moving like "spooky actions" across contexts. As illustrated below, I draw on data to redefine how we understand quantum entanglements across the sciences by teasing apart the possibilities and complications of our intra-actions. But before I can fully explain quantum entanglements, I describe a few foundations of quantum mechanics, or how the smallest parts of the universe act. While there are several prominent figures that have contributed to the field of quantum mechanics, I focus here on Niels Bohr, Albert Einstein, Erwin Schrödinger, and John Stewart Bell.

Entangled Cats

The Copenhagen interpretation of quantum mechanics, a theory explicated by Niels Bohr and Werner Heisenberg, was one of the first attempts to understand the quantum world. The Copenhagen interpretation argues that systems do not have definite properties prior to being measured. What this means is that matter exists in states of "superposition:" an object can be in multiple states at once, which are completely indistinguishable until the object is measured. One popular example of the Copenhagen interpretation is the thought experiment known as "Schrödinger's cat", developed by Erwin Schrödinger. This experiment explained how a superposition might function, teasing out how a particle can be in two or more states at the same time when not observed. In the experiment, Schrödinger morbidly proposed placing a cat in a box with poisonous gas. The gas has a 50% chance of spilling and killing the cat.

However, until someone looks in the box, there is no way to tell if the cat is dead or alive. Repeat the experiment enough times, and half of the time the cat dies and the other half the cat lives. The quantum mechanical interpretation is that the cat is actually in a state of superposition, both dead and alive at the same time, and that it is the act of looking that forces nature to make a decision. In other words, our curiosity can literally kill the cat. One can argue that the binary proposed here is incomplete because if the cat is found dead, then it was alive, then dying, then dead, but was never dead and alive at the same time. The metaphor also falls apart because the cat is not a quantum particle with infinite states of being and because, as queer theorists have noted, binaries like alive/dead or here/gone tend to be false constructions of bodies and how they move (e.g., Johnson, 2016).

A second example, perhaps more everyday and less malicious than murdering cats, might be to imagine walking down the hall of a school while a student from a few years prior approaches you to say hello. The only issue is that you have forgotten this student's name. Quickly you search your memory: Was her name Imani? Shanice? Alyiah? She is getting closer, so you go out on a limb: "Hi....Kiara....?" In this case, you are correct; the student's name is "Kiara" and she happily greets you, her favorite teacher. However, the collapsing of possible names is only part of the equation in this example. One of the reasons that Schrödinger's cat is limiting is because theorists tend to focus on the cat; seldom are questions explored about who opened the box. This means that we do not know things like: How many people are present who might look in the box? Is the person who opened the box the person who looked? What is that person's relationship with cats in general or with the specific cat in the box? Does the poisonous gas impact the person opening the box?

So, rewinding the scenario a bit: you are walking down a hall and you see a student who you believe recognizes you. You have no way of knowing what the student is thinking. Perhaps she does not recognize you, but you recognize her and, as a result, you start searching your memory for her name. Perhaps she had no intention of saying hello to you and was just waving politely. While we might not know about the student's side of the equation, much like we do not know about the person unsealing the box, what we do know is that ways of being are often coconstitutive (e.g., Barad, 2007; Hultman & Lenz-Taguchi, 2010). Although some might have more force than others, like the teacher whose power is certainly present in the context, the entanglement cannot be ignored.

One can think of the moment where a teacher is guessing at names in her head similar to the measurement of a wave function. A wave function simply describes the probability of finding the state or location of a particle, whether it be the particle's spin, position, or momentum. In that moment of panic, the name of the student can ebb and flow, like a wave, until one name is revealed. When the teacher does not know the student's name, the state of that name exists across multiple possibilities and can be understood as a superposition. When the correct name becomes known, this is similar to what physicists call "collapsing a wave function", wherein the infinite possibilities of how a wave might function collapse into a singular possibility, revealing the state the system was actually in prior to being measured by calling out the student's name.

This is where the analogy breaks down. The reason a teacher might not recall a student's name stems from a lack of memory, not from the student having infinite names. A particle, on the other hand, is very different. Understanding the properties or location of a wave is not a matter of lack of information on the part of a person conducting the measurement, as it is with student names. This is a fundamental property of quantum objects: their state of being really *is* spread out and only upon measurement do these objects collapse into the state that is observed.

Identities can also be understood to exist in a state of superposition, a multiplicity of possibilities and ways of being that are often forcibly collapsed when they encounter sociocultural norms and values in much the same way that measurement collapses particles from infinite to binary possibilities. Unlike the example above about recalling names in the hallway, identities *can* be understood to exist in a state of superposition, much like quantum objects. However, collapsing the wave of identities does not remove the spectrum of who one is or how they can "be" in the world. The collapse of one's superposition of identities can therefore also be understood as a function of agentic contingencies, a pushing against one's agentic possibilities until the wave function collapses, reifying sociocultural norms and values. This is not unlike anthropological dialogues about how identity (e.g., Varenne & McDermott, 1998), possibly existing in infinite ways of being, is often collapsed into a finite way of being when a person is identified by others. Thinking of one's ways of being, knowing, and doing as a superposition, however, attends more closely to the multiplicities that exist in one's identity as it is, was, and might yet be—whether or not it collapses upon "measurement".

The aforementioned collapse is the result of intra-actions, in which one's agency is constructed through relationships. While Barad primarily discusses intra-actions as productive in their creation of agency, it is also important to note how agency is limited by other forms of power and how in the construction of agency power still limits one's possibilities. If queerness presents a set of possibilities, cis-heteronormativity imparts limitations to one's ability to move within sexual orientations, or one's agency. One could also argue that genderqueer bodies are in a superposition of gender or that bisexual and pansexual bodies are in a superposition of sexual orientation, which collapse in moments where they are voluntarily or forcibly measured.

More Than an Entangled Feeling: Quantum Entanglements

To this point, I have discussed superpositions as they might exist independently, becoming entangled with agentic contingencies. Specifically, I have examined how multiplicities are contextualized within infinite ecological layers through the act of their examination, often collapsing into binaries. As significant as this is, it is equally important to consider what happens when two particles interact so much that their states become interdependent. In the examples of Schrödinger's cat or the student in the hall, physicists consider the superposition independent, collapsed by the measurement. However, one could also argue that the outcome was dependent on the measurement, that the collapse of the wave function, the superposition prior to collapse, and the measurement are all entangled in some way—always already in relation. This is not the same as quantum entanglement, or the idea that *interaction changes a particle* and how it reacts to measurement, which is another form of intra-action.

Where "entanglement" on its own refers to the deeply connected nature of bodies, quantum entanglement is different. In it, particles are so deeply entwined that one can change how the other moves over distance, well after being separated. Returning to Schrödinger's morbid thought experiment, imagine if this exercise was repeated with two cats in two boxes. Only these are not regular kittens; these are quantum-entangled kittens. Both cats exist in their own boxes in a state of superposition but, because of the entanglement, their fates are connected: when one is measured, the other changes to reflect that measurement. The distant particles (kittens) are somehow influencing each other at a rate faster than the speed of light. If you open a box to find one cat alive, the cat in the other box will automatically be dead because of the first cat's fate.[2] The sudden change in entangled particles is what Einstein called "spooky action at a distance;" it is also called the EPR paradox after the physicists (Einstein, Podolsky, and Rosen) who proposed the experiment. In the EPR paradox, when particles become entangled, they impact each other, changing the properties of both. When measuring one, the other should change, regardless of the distance. Einstein argued that this was impossible because it would break relativity, or the idea that events are deterministic and that cause will always match effect.

It was not until 1964 that John Stewart Bell devised what is known as Bell's Inequality and, through a set of mathematical equations, demonstrated that quantum mechanics and entanglement do exist. Einstein was incorrect; the EPR paradox was possible. In the quantum world, atoms, electrons, protons, and neutrons do behave like particles, but they also move like waves. In other words, matter is similar to light or sound in that it has wavelike properties. If an environment is calm, waves follow a simple reciprocating motion, up and down like a ripple of water. However, if the wind is blowing, the wave receives interference, and what was once an evenly spaced motion can become incoherent and erratic. This is not unlike how a person's affect is subject to local and less local events, ideas, and histories in ways that impact the wavelength through which one moves in the world.

While quantum entanglements have been documented in controlled lab settings, these experiments have generally been limited by factors such as temperature, size, and distance. The lab setting has to be cooler than typical room temperature, the size of the objects being entangled must be microscopic, and the distance of entanglements is difficult to measure. However, in 2011, Lee et al. were able to entangle the vibrational states of millimeter-sized diamonds at room temperature. In 2014, David Kaiser and his colleagues managed to isolate particles well enough to entangle and then measure them across various distances. In short, the work of these contemporary physicists has proved the EPR paradox and Bell's inequality to be correct. Perhaps more significantly, these studies present several possibilities regarding what other bodies might be entangled with, how those entanglements can be measured, and what this might mean for literal and metaphorical intraconnections around the world. The next section picks up these possibilities to illustrate how bodies became intraconnected through an ordinary affect—hope—at a middle school talent show.

Entanglements of Hope

New Horizons is a small middle school in a midsized city on the East Coast of the United States. At the time of this study, there were 162 students at the school; 88% identified as African American, 10% as Latin@/x, and 2%

as white. Additionally, 90% of the students were identified as coming from low-income households. Of the approximately 15 teachers, two identified as faculty of color. Although New Horizons was a charter school, it was located in a state where charter schools functioned differently than in other places across the country, as they were connected to the school district. The school's faculty were district employees and part of the teachers union. Funding came partially from the district, supplemented by the business supporting the charter—in this case, a nonprofit organization whose mission was to "disrupt the cycle of poverty" through education.

Like all charters across the city, students were admitted to New Horizons through an application process within a citywide lottery system. While parents and teachers noted that New Horizons was perhaps not the most academically competitive middle school in the district, the school's administration boasted that 15–20% of their students scored proficient in reading and math and that many alumni matriculated to the "better" high schools in the city. On top of the discrepancy of parent and teacher perspectives from those of administration, students frequently articulated that they were not always enamored with the faculty, school facilities, or their education at New Horizons. Nonetheless, students generally identified the school as, to quote one student, "a hell of a lot better than the school near home".

When I had arrived at New Horizons, I was provided with a list of students who were identified as strong candidates to participate in the group. This was because teaches' had worked to identify which students were potentially queer. I decided that my first task at the school was to dismantle the original list. From my perspective as a researcher and former K-12 teacher, forcing students to participate in an activity because they had been identified as queer by others was not ethical. Additionally, I was curious as to how many LGBTQ+ students were absent from the list that might have a desire to participate. I asked to put out a schoolwide call for those interested in thinking about questions of identity in relation to sexual orientation and gender at the intersection of race, ability, and home language. This invitation garnered a group of 14 students across sixth, seventh, and eighth grades. Students quickly chose the name "The Blue Dragons" for their group; they felt that the name represented their "spirit animal" because dragons were "cool but still able to spit fire at any haters".

While some students in the group came out to me over the course of the two years that I ran it, I never asked them about their developing identities. Instead, I focused on giving them the creative space to explore such topics on their own terms. While there are certainly arguments to be made that it might have been my ethical imperative to specifically address questions of *their* identities, I decided to give them the space to talk to me only if and when they decided to do so because they were already overwhelmed with their peers' and some teachers' questions. As a result, almost all of the students did come out and, over time, all of them asked questions about sexual orientation or gender identity. My decision empowered students with the space to speak as they wanted, on their own time, and at their own comfort level. Based on what the students did tell me, of the 14 "dragons" who participated, two identified as Latinx, one as white, two as multiracial, and nine as African American. In terms of gender identity and expression, three discussed being genderqueer. Regarding sexual orientation, nine came out as either lesbian, bisexual, gay, or questioning.

Although the students were diverse in race, gender, and sexual orientation, there was one common thread that most shared: many articulated being categorized as "a problem" by adults in the school. The male-identifying students in the group often explained that they were considered a problem in school and at home, always being yelled at for not "getting things right". As one male-identifying student informed me, "I get in trouble at school. I get in trouble at home. I get in trouble walking home for not joining up with the dealers. When exactly am I *not* in trouble?" In contrast, female-identifying students described their "difficult" status differently: on one hand, this status was because they struggled with their grades, and on the other, because their parents and guardians had to "fuss over" their futures. For the eighth graders, this often meant that their parents and guardians were concerned with their developing sexuality. As one student, Rashida, said, "My mom doesn't want me to be a lesbian, but she doesn't want me to get pregnant either. If she would just accept that I'm gay, she could stop losing her shit over the pregnancy part". For this group of students who were frequently identified as "trouble", they often described their time with the group as being where "all the trouble got together to get stuff done".

Genderqueer students described themselves as "troubling" across contexts. At school, they expressed consternation at constantly being asked questions about their gender identity, particularly when such inquiries were related to other students wondering if they would eventually have gender-confirming surgery. One student, Dominique, spoke about their parents' difficulty with accepting their nonbinary identity at home, noting, "it's not a problem for *me,* it's a problem for *them.* Why should I have to be one way or another just for my parents? I just want to be the me I feel in the moment. I don't get why it's an issue". Students who were fluid in their sexuality explained similar issues, especially when they "decided not to be gay" or straight. For girls, this was often marked by other students saying that they were "just trying to get attention", while for boys it was often labeled by peers as "a decision to be gross" when they pursued gay relationships. After all, while any middle school student might consider engaging in sexual behavior regardless of their sexual orientation, gender identity or expression, race, home language, or other factors, students can also be shy and sexuality can grow out of quiet beginnings, like holding hands, crushes, or the ever-anticipated first kiss.

During our first few meetings, I provided suggestions for the direction and purpose of the group but tried to allow students to shape these as much as possible. Similar to allowing students the time and space to discuss their own identities, I wanted to give them the space to decide how the group would be run and what, if any, projects they wanted to pursue. After meeting with the lead teacher, Bridget, and the principal, it was clear that while students had several extracurricular options at New Horizons, they might feel as though they lacked the ability to decide how these groups were run. Because my purpose in being there was to think about developing identities and to provide a space for students to think about this with someone else, I felt ethically compelled to use my available privilege as the professor in the room to open up spaces of possibility, to open the possibility of freedom to "be" and "become" during our time together each week.

The first project that the students suggested was to start a podcast. They were intentional in their decision to embark on this project, explaining that podcast production would serve two purposes: First, and likely foremost, some of them wanted to learn how to mix sound out of interest in future career possibilities in the music industry. Second, students voiced concern that there were too few spaces available to explore identities as they were related to local and less local communities. In other words, they expressed the tension of peers and adults asking about their sexual and gender identities as if they were siloed, rather than imbricated with other people and groups. When such factors were addressed, it was usually done to investigate "bad behaviors" like smoking or cheating on a test. The youth in the group therefore felt that a series of interviews would be a great way to think about their respective identities as they were formed in relationship to teachers, administration, fellow students, and the surrounding community. Students decided on topics, wrote the questions, conducted interviews, and edited the audio files primarily on their own.[3]

At the end of the first year, the group was invited to present the podcasts to parents and guardians at a talent show meant to highlight the work of students participating in school-sanctioned extracurriculars. These ranged from choir to finance club, from African drumming to improve club, from ballet to our Blue Dragons. Prior to this moment, aside from learning more than I imagined about podcasts from the students, I had been analyzing data gathered in the group and theorizing what I call "assemblages of violence" (the focus of the next chapter). In its foundational, basic construction, my understanding of assemblage in this case leaned on the related yet distinct frameworks of Karen Barad (2007), Alexander Weheliye (2014), and Jasbir Puar (2007, 2013), who theorize assemblages as impossibly intertwined sets of relations from which no single thing can ever be extricated. Specific to my theorization of an assemblage of violence was the ways in which violence is always entangled with other aspects of life—so I entered the talent show expecting more violence from the audience.

There I was at the talent show, thinking about violence—the violence of the school toward LGBTQ+ students, the violence against queer youth and youth of color found across school systems, and the shame that is all too often normalized and weaponized as a mechanism of control against marginalized people and groups. That's when it happened. In the midst of tired parents, younger siblings crying and playing in the back, and students shifting nervously, the lights dimmed, and the first group broke into the song *Seasons of Love*: "525,600 minutes, 525,000 moments so dear..." (Larson, 1996). Looking around, I realized that everyone had stopped. Even the babies on the edges of their parents' knees were quiet. The whole hour-and-a-half program went like that, with everyone listening and everyone loving.

Afterward, I stopped several Blue Dragons' guardians, asking them what they were feeling during the show. While I had expected to hear the word "pride", the word I heard most was "hope". Parents and guardians said they felt "hope for their children's futures". When students described the experience, they too used the word "hope", yet their hope was feeling hopeful that the experience would allow people to see them as "more than gay" or "more than a problem". Hope, in this case and in most cases, was fundamentally different for different people, though its expression could be read as the same.

For some time after that moment, I felt haunted. Haunted by hope. This was not because there is anything wrong with hope, but because experiencing hope in that moment felt jarring. I had incorrectly assumed that there would be pushback when a group of queer youth stood up in front of their peers and community. While there was plenty of resistance to queerness in the school and broader community, in that moment, homo- and transphobia seemed to erode slightly, leaving behind a glimmer of hope and joy. I use this haunting to think about how hope can be traced. What is the trajectory of hope within an assemblage of violence, and what does this mean for LGBTQ+ people and groups?

Following the talent show, several events occurred. First, the hope that stemmed from the event seemed impossibly entangled with the affects of other students, school administrators, and guardians in two keyways. This hope seemed to generate and sustain other forms of hope. It was more than infectious; the levels of hope seemed to rise without contact between bodies. For example, in discussing the talent show with a guardian whose child was not in the Blue Dragons, she said she had been feeling "inspired" since the show in ways that she could not describe. When I asked if she had been to the school since, she explained that she had not. Further, until that day, she had had limited contact with her own child, as he was staying at his father's house. This parent said that it didn't feel like it was only the show that she was feeling, but rather that the feeling was "in the air". Other students similarly described this positive feeling as something that could not be named, yet, was palpable during the evening of the talent show.

While this hope hung in the air for some students, for others, hope receded and faded into despair as everyday racisms and homo- and transphobias pushed against their (re)newed senses of self from the talent show. Some, like Davonte, explained that they "feel more sad inside". What is important here is understanding the always-already-connected nature of hope and despair, which are often falsely presented as dualities. Participants were not culturally duped (Hall, 1973), rather they were constantly reaching for false dualities to better understand their own lives, as so many people tend to do. One only needs to recall the constant tension of racism or homophobia where, despite conditions of oppression, hope can arise—only to be tempered again by despair and hatred.

Unlike how hope is often theorized as independent of or in duality with despair, I attend here to how hope acts within its quantum entanglements with violence as it moves through, within, and between bodies. This includes the bodies of the students as well as local and less local cultural bodies. At the talent show, hope entangled with other forms of hope. As these feelings of hope slowly receded because of the school structure and systems of schooling, others who had experienced hope were similarly impacted. Conceptualizing hope as an aspect of quantum entanglement means not only that hope has trajectories that are agential in the ways they shift and change the bodies with which they intra-act, but also that, because of such entanglements, hope can be a flash of light born from spooky movements from beyond the grave within a violent assemblage. This does not mean that hope is always part of a quantum entanglement; instances of hope can and do exist as complicated, interwoven understandings that are not so deeply entangled that one person's hope is changed or informed by someone else's across space and time.

At the talent show, the room was filled with straight kids as well as queer youth, white students as well as students of color. The spooky action of hope applied to all of these students. Intra-actions are inherently unpredictable. Understanding the quantum entanglements within an assemblage allows for a degree of predictability within an inherent sense of unpredictability. Sedgwick (1993) discusses the indelible reality of queer youths' lives as tied to questions of self-harm and suicide because these are normalized through cis-hetero culture. In this discussion, Sedgwick argues that homo- and transphobia are business-as-usual, a point that is central to the students at New Horizons. Given the current and consistent raise in queer youth suicide, perhaps a bit more attention to spooky actions of affects like hope as they occur across spaces, times, and places should become our new irregular-regular reality.

Analyzing Entanglements

Hope can be understood as existing within, rather than separate from, assemblages of violence. The "stickiness" of affects, as Ahmed (2010) argues, is often how affects like hope are theorized. Hope, however, goes beyond the stickiness of affect, where affects simply stick to and through people, extending to the entanglement of intra-actions where one's affect seems to impact others' ways of being across distances. One way to theorize the complicated nature of hope might be through Berlant's (2011) discussion of cruel optimism, or another might be through Ahmed's (2010) take on happiness as consistently conditional. Both scholars argue to some degree that because of sociopolitical and cultural conditions, people seek out happiness and joy through outlets that will eventually lead to despair. A salient example is Berlant's discussion of food, where people tend to seek out food that is satisfying but ultimately unhealthy due to its content or high caloric count, an idea to which I will return later in this chapter.

What makes assemblages of violence so insidious is that the possibility of hope exists but is not always realized. There are at least two ways to understand this presentation of hope. One rather pessimistic understanding is that you can never be fully hopeful. Another is to be mindful that hope exists within a context of violence that is often beyond one's control, but that experiencing hope and joy resonates against violent intensities. Within the assemblage of violence at New Horizons, hope felt unpredictable. I didn't know that hope would propagate during the talent show any more than I knew how or when violence would reterritorialize into despair. What I did know is that hope and despair seemed to be intraconnected between bodies and through their intra-actions. Although hope and despair are commonly two entangled aspects of assemblages of violence, there are other facets including joy, happiness, and inattention. These are not just an "entanglement" as the term is often used to discuss the agency of things or through new materialist dialogues, but rather, as Barad reminds us, these are entangled at the quantum level; they are quantum entanglements.

Although Einstein's phrase "spooky action at a distance" was initially used to jest at the idea that particles could be inextricably linked, particles do in fact become so deeply entwined that they can influence each other across time and space. However, there is no way to predict which particles will become attracted and entangled, just as there is no outside of such entanglements. Returning to the talent show, hope acted as an agent among bodies such that it was impossible to predict how the various entanglements would intra-act. Taking my own experience as an example, I did not anticipate that the results of what could have been an oppressive event for queer and questioning youth, standing in front of their classmates and community to present their work would be transformed in ways that I had not begun to consider. This happened through what I call the agency of hope.

In the case of entangled hope, measuring one person's hope does not necessarily mean that another person automatically felt despair, although there were certainly cases spurred by the talent show in which hope and despair were deeply linked. Of course, affects are not quantifiable. (This is why the pain scale used at doctor's offices is often archaic and not necessarily helpful to a diagnosis.) In "measuring" affects, measurement is more about considering the manifestations of the affect as it is related to and across contexts than about a traditional measurement in terms of quantifiable information.

As agentic contingencies of the sociopolitical system in general, and systems of education more specifically, limited students' ability to maintain hope, the fatigue of lost hope also spread among those who shared the initial emotion. Perhaps this is because the superposition of entangled beings is not always in inverse relationship, but rather that the measurement of one body can have a direct correlation on its entangled other. In other words, when teachers worked to decide what the hope was, what it meant, and to whom it was related, it impacted all students engaged in the entanglement. Or, perhaps, given that there were many entangled bodies in the room and the school, students fell back into despair as their labels of "trouble" engulfed feelings of hope, while the children who were labeled "good students" were able to regain their social capital.

The labeling of some students as "trouble" often resulted in individuals becoming more closeted about their identities in order to better negotiate cis-heteropatriarchal ideas and ideals. In other cases, such labels became a self-fulfilling prophecy. There are many ways to interpret being in "trouble" in a school, from students who create trouble in order to exercise agency and leave class to a teacher creating a "troubled" narrative to enact control over students. In this case, trouble speaks to the broader conditions of the school and community but not necessarily to the act of being in or deciding to engage in acts that were considered "trouble".

While Barad uses quantum entanglements to think about how intra-actions produce agency, here I focus on spooky movements. In the EPR paradox, bodies that are linked can instantaneously influence one another, regardless of distance, perceived relationship, or any other such factor. In the next section, I explore how "bodies" are defined, made, and unmade through intra-action. Specifically, I pay attention to how bodies—human and non-human alike—are entangled at a quantum level. In Prisha's narrative, which I recount below, it is cultural bodies—both local and less local—that are entangled with the human body, each shaping the agency and understandings of the other in the process. Or, as anthropologists like Varenne and McDermott (1998) argue, identity is both a personal self-determination and dependent on how others identify an individual. Adding another layer, when one entangles with cultural bodies, these form and inform one's agency and sense of self. Stated plainly, what makes a spooky movement spooky is that shit comes out of nowhere. Not only does an entanglement shape the possibilities of both particles in their measurement, prior to and during the entanglement, particles are moved as much as they are moving. Physicists have theorized that such entanglements can even survive their deterritorialization, or, as Emily Brontë (1846) may have written in *Wuthering Heights*, a quantum communication might equal a flash of light from beyond the grave.

आशा: "Aasha" Means Hope

Aasha Women's Shelter is housed in a large, concrete building. The exterior is neatly kept, with a variety of plants surrounded by the rust-colored dirt that is common in the region. A narrow stone path leads from the gate to the building, and when visitors make their journey from their cars, a small group of dogs barks and sniffs excitedly. The initial seating room is small, outfitted with only a desk, a few chairs, some artwork, and an old fan. Although the dogs provide an aural form of protection, the gatekeepers are clearly the center's directors, Neetu and Deepti, who take the responsibility of their jobs seriously. When we arrived for the first time, Neetu stood in the doorway between the sitting room and the courtyard, marking the line between public space and the residents' living area. The sitting room opens into a square-shaped courtyard. Residents' rooms, a kitchen, and a common space with couches and a TV made up the perimeter of the open space. This common space was well maintained and, as Neetu explained, each resident was responsible for helping with chores to sustain the "order of the place".

The residents of Aasha ranged from ages 30–55. The shelter was predominantly funded through donations from local religious organizations, which supplemented the minimal funding granted by the state. The state provided funding to the shelter not because it functioned as a reprieve for survivors of intimate partner violence but because there was a significant population of residents who were considered, in the words of Neetu, "undesirable by their families" due to a range of disabilities. Survivors of violence stayed at Aasha either as permanent residents or temporarily while the court sorted out responsibility for the violent actions taken against the women and, too often, their children. This meant that if the court sided with the husband, such as by finding the wife "negligent" in her home-making duties, she might be sent back home. If found culpable for his actions, the husband would be issued anything from a warning to a fine or, in rare cases, would be incarcerated for some time, depending on the severity of his aggression.

Prisha was considered young among the shelter residents. When she arrived at Aasha, she was just 24. Her son was five. Because boys were permitted to stay at the shelter while they were children but not through adulthood, Prisha was clear that she needed to rebuild her life as quickly as possible before her child was too old to stay with her. However, Prisha, like many of the girls from her small village in southeast India, had not cultivated the skills necessary to earn enough income to live independently as a single mother. While her son, Arjun, attended school during the day, Prisha took the opportunity to learn skills like basket weaving at the shelter. She also learned less tangible skills like how to handle finances. Her decision to leave her husband had put her under social strain with her family and the surrounding community. Her need for financial literacy and additional skills thus stemmed from the pressure of being able to integrate into a new community as an independent mother.

Despite this pressure, Prisha stayed strong. She explained that she had already lived through so much that "one more thing could not break" her. The unpredictable beatings, being raped to satisfy her husband's sexual desires, and the dismissive responses of her family had already created several points of friction on Prisha's body and emotions. Then there was the act of getting out. Similar to many conservative places in India, in the area where Prisha lived there was a growing network of women who functioned like an underground railroad for victims of domestic violence. For Prisha, it was the संरक्षक, *sanrakshak* (guardians) who held three important roles in hers and surrounding villages. First, they worked as liaisons between women and the court system, negotiating on women's behalf. Second, they were connected to forms of physical protection; Prisha explained that the sanrakshak were known to surround a batterer in the street and beat him, leaving bruises as a social reminder of his actions behind closed doors. Finally, when all else had or was likely to fail, they helped move women and their children to places like Aasha Shelter, or to private homes, to offer reprieve.

This last option was not to be taken lightly, as it generally meant that the woman could not easily return home. Yet Prisha had used the sanrakshak's network to move several states away from her abuser and find peace at Aasha. Over time, Prisha became a rather skilled artist in weaving and her confidence grew. Arjun also showed signs of healing while living at Aasha. Neetu, one of the shelter's directors, noted that when Arjun had arrived, he was a shy boy who hardly spoke. Skittish and constantly alert, Arjun spent more time at the hem of his mother's sari than he did playing with other children or exploring the shelter's compound. Over time, Arjun seemed to heal, making friends at school and bothering the guard dogs who roamed the shelter grounds.

Despite Aasha offering a space of reprieve for residents like Prisha, the shelter does not exist in a vacuum. Neetu and Deepti explained that over the years they had faced violence from husbands and families looking to retrieve their loved ones. Prisha noted that even if her own family never came after her, the incidents of men shouting angrily outside the compound served as a stark reminder of her own abusive context. In these cases, the shelter called the authorities, and the unwanted visitors were removed from the premises. In addition, women received newspapers at the shelter and could read about abuse cases that made national headlines; the *Times of India*, for example, consistently reports cases of rape and domestic abuse. One case in particular was covered by the *Times* a few months after my initial visit, which Deepti recalled impacting residents. The article describes a woman who, after her family being unable to pay her full dowry, was dragged to the roof of her house by her husband, "tied up with a plastic rope and set on fire after he emptied a five-litre kerosene canister on her" (Chauhan, 2016, p. 2). According to Prisha, these cases of extreme depravity served as a reminder that if "those were the headlines, imagine what didn't make the paper".

After a few years, Neetu and Deepti recommended Prisha to meet a young man who lived near the shelter. Speaking with the confident tone of a woman whose calling has been to protect other women and children, Deepti explained that she "knew this man would be good for Prisha". Before long, Prisha married this young suitor and they moved together to Mumbai to follow his career. Deepti and Neetu kept one of Prisha's weavings to proudly display in their waiting area, noting that it was the "gift of a success case" to the shelter. Sometime after my initial visit, I asked Neetu if she still heard from Prisha. She said that occasionally Prisha writes with updates, noting that she does not need to work because her husband makes enough for her and Arjun to "live a good life, including a second baby" that had been born by the time the last update was sent. Deepti noted that as "divorce rates continue to climb across the country, it was nice to see the shelter being successful in cases of happy marriage".

Cultural Entanglements

There are many ways to interpret Prisha's narrative. A pessimistic understanding might be to note that even after learning independence, she fell back into familiar patterns of patriarchal protection by getting married and following her second husband's job. Another read might be that she found a nonviolent partner as a way of prioritizing the physical and emotional safety of herself and her son in a country where the divorce rate was less than 1% at the time of this study (Biswas, 2016). My goal is not to make psychological judgments about Prisha's decision to remarry, especially as there are too many possible reasons for her actions that cannot possibly be analyzed here. Instead, I use her story to consider how we are connected to cultural norms and values.

Anthropologists and sociologists have engaged in long dialogues regarding the relationships between, through, and within people and cultures (e.g., Geertz, 1973; Jackson, 1968; Ortner, 2006). Significant to Prisha's narrative is how cultural bodies intra-act and are enmeshed with people. Following Barad's description of intra-actions, agency is coconstituted among both human and nonhuman bodies during an event. It is also possible for bodies to become entangled during such intra-actions, much in the way that quantum entanglements form and inform ways of being, knowing, and doing (Agar, 2004; Trnka & Lorencova, 2016).

Because culture is pervasive, human bodies are in constant intra-action with its norms and values. Culture here is its own agent—like racism or hope—moving people as much as it is moved by us (Hall, 1984). Additionally, as there is no way to know when or how bodies might become entangled at the quantum level, it is possible to say that we are all in quantum entanglement with culture at some level. Like a quantum entanglement, it is difficult to describe human bodies independent of broader cultural bodies. Human bodies also intra-act physically and emotionally with cultural ideas and ideals. In Prisha's case, the normalization of violence against women consistently intra-acted with her ways of being, knowing, and doing. This was the case from her presence at a shelter necessitated by her abuse, to the men who tried to intimidate other residents into leaving the shelter, to the news articles that described violence but did not always admonish the collective cultural cycles of violence or the actions of the abusers.

As the quantum entanglement of survivors and abusers is measured—either by the court, a woman's local community, or some combination of the two—both the body of culture and the body of the survivor continue to intra-act, despite distance. In Prisha's case, she had moved several states away but remained connected not only to her local culture but also to broader cultural values that acted as a web between cultural understandings across the country. For example, Deepti was clear that Prisha's marriage was a sign of "success" against what she perceived as a rising tide of divorce in the country. Despite her position as the center's codirector, Deepti still spoke traditionally about marriage and, in some cases, about her agreement with the idea that it is a woman's role to please her husband. It is as if Deepti's measurement of marriage as a nationally failing institution was central to Prisha's decision to date and eventually marry.

This quantum connection is important for several reasons. First, as women continue to fight for their rights nationally, it is significant to consider the local impacts of such entanglements on the bodies of women and children. For example, when women formed a human chain in the state of Kerala to support gender equality in 2019 (BBC, 2019), participation in men's rights organizations was also surging across the country in reaction to gender equity movements (Cockerill, 2019). Neetu described these imbalances like a seesaw: when one side gained power, someone else lost somewhere. "Men get nervous", she said. "When women start to fight nationally, some think they need more control in their home".

Second, as marginalized bodies continue to exist under limiting structures, quantum entanglements mean that the inverse can also be true. Just as described in the talent show, hope can engender hope. As the #MeToo movement has grown into a global phenomenon, women in countries like India have been empowered to remove themselves from abusive contexts or to form a nationally recognized chain of support for gender equality. In Prisha's case, her ability to hope for a more stable future was deeply entangled with the cultural measurement of women's rights and ideals of marriage. Although she learned skills to allow her to be relatively independent, she remained connected to broader norms. Aasha, the pseudonym I've given to this shelter, means hope. But while the place brought reprieve, it did not always bring hope. Women like Prisha remain connected to ideals of and paths toward hope that are culturally constructed. Or, as Prisha noted, "Hope for my son means showing him how to *be* and what a (nice) family is, right? That's hope enough for me".

Limitations, Possibilities, and Entanglements

The emergence of cultural norms and values is possible because of the quantum entanglement of bodies (Trnka & Lorencova, 2016). But a limitation of thinking with quantum entanglements is that contemporary literature describes them as resulting in opposite reactions between entangled bodies when they are measured. Remember: one dead kitty, one live kitty. However, it is important to remember that scientific communities are still discovering the conditions under which quantum entanglement can occur, how long these relationships last, and what they mean in terms of how one body impacts another. The entanglements described here necessitate a more complex understanding of what it means when bodies are so deeply linked that they impact each other across time, space, and place.

Take, for example, recent research on how birds navigate during migration (Dare & Fletcher, 2019; Vedral, 2011). Ornithologists have theorized that birds use a quantum entanglement bonded through a protein in their eyes that helps them respond to Earth's magnetic field, guiding migration (Dare & Fletcher, 2019). Quantum entanglements had previously been observed in controlled laboratory conditions that were usually cool, dry, and precise. Birds' eyes are warm, wet, and comparatively chaotic.

This departure from traditional understandings of quantum entanglement is significant to the narratives I have presented in this chapter, as perhaps the measurement of an entanglement does not have to result in opposite reactions. This move away from traditional interpretations of quantum entanglement is a deliberate push against the binaries that are normalized within quantum mechanics. In Prisha's case, her entanglement with culture meant that as her sense of hope increased, the agentic contingencies of culture pushed against this hope. The measurement of her increased hope can be understood to have impacted broader cultural norms that work to maintain the status quo. Similarly, if her hope had affectively entangled with the hope of other women at the shelter, like happened during the talent show at New Horizons, hope might have spread. In this case, measuring Prisha's affect and the entangled affect of another survivor at the shelter would have had a positive correlation.

In this way, the notion of measurement might be considered positive. From a perspective of equity and access, measuring hope or changes in people's ability to feel and express hope, as well as carefully considering the causes of such affects, might change the dialogue from "it happened" to "here are the pathways that allowed it to happen". This illustrates the tension between the method of tracing hope and the metaphor of quantum mechanics; measuring affects is difficult if not impossible. But in assessing changes in affect, one can consider not only how to disrupt violence but also, perhaps, how to propagate affects like hope and joy.

More directly, I call for a more complex understanding of the results when bodies intra-act through a quantum entanglement: a "both/and" rather than an "either/or", a move away from the reproduction of binaries traditionally understood within quantum mechanics. Instead, I argue for attention to the deeply linked bodies that impact and intra-act across distances, resulting sometimes in affects that buttress each other, sometimes in deterritorialization of one or both within the entangled pair (Deleuze & Guattari, 1980; Trnka & Lorencova, 2016), and sometimes sliding between these sets of possibilities.

The inability to predict associations between entangled bodies upon measurement might be due in part to the complementarity principle. In 1928, Bohr argued that it is impossible to observe both wave and particle properties of atomic dimensions at the same time. Bohr explained that what is observed is highly dependent on the experimental arrangement. For example, if one is looking for a particle's movement, they will observe the movement but not the position. This is akin to anthropologists' understanding that one can never be in all places and see all things despite, as Wolcott (1994) argues, the desire not to "get it all wrong" (p. 356). Perhaps in looking for violence and finding hope, I missed the negative impacts of hope on the body, or by hearing how hope was tied to broader cultural violence, I missed how hope was also maintained through such values.

Again, I return to the agentic contingencies that limit bodies within quantum entanglements. At New Horizons Middle School, such limitations were visible in the collapse of students' identities, recast here as superpositions. On one hand, their queerness was identified for them, and on the other, they were identified as trouble across multiple scales. In Prisha's case, she was limited by her entanglement with marginalizing cultural norms, as her future became further knotted within her second marriage. Much like humanist descriptions of sociocultural patterns and boundaries, these norms bind one to prescribed ideas and ideals as much as they can enable change and mobility.

Within an assemblage of violence, the movements of entanglements are of critical importance. As Einstein argued, quantum entanglements exist as "spooky actions at a distance".

The movement of entangled pairs and the measurement of these pairs are important to understanding an overall assemblage and its impact in and across bodies. As assemblages tend to burble up and recede in both expected and unintended places, the movement of one entanglement remains connected to others' ebbs and flows.

Notes

1 I use "animal" to denote and discuss the relationship between "humans" and the "animal world" as inseparable, albeit still steeped in questions of power. However, given that the participants in this book have experienced oppression based on race, gender (with its multiple expressions and identities), and sexual orientation, it is crucial to note that the use of "animal" or "human animals" is not a turn toward oppressive histories where the oppressed were considered less than human. Rather, it is an attention to how we as humans are inseparable from the animal world in the era known as the Anthropocene.

2 Again, the entangled kittens set up a binary that does not apply well to how other bodies act and intra-act.

3 You can listen to students' podcasts here: https://soundcloud.com/user-299695014

References

Agar, M. (2004). We have met the other and we're all nonlinear: Ethnography as a nonlinear dynamic system. *Complexity*, *10*(2), 16–24.

Ahmed, S. (2010). Happy objects. In M. Gregg, & G. J. Seigworth (Eds.), *The affect theory reader* (pp. 29–51)). Durham, NC: Duke University Press.

Anderson, G. (2019). *Strange coincidences in your life. Small curious events. Forebodings. Telepathy. Does it happen to you too? Quantum physics and the theory of synchronicity explain extrasensory phenomena*. Italy: Bruno Del Medico.

Barad, K. (2007). *Meeting the universe halfway: Quantum physics and the entanglement of matter and meaning*. Durham, NC: Duke University Press.

Berlant, L. (2011). *Cruel optimism*. Bedford, O., & Hwang Durham, NC: Duke University Press.

Biswas, S. (2016). *What divorce and separation tell us about modern India. The British Broadcasting Corporation, Retrieved from*: https://www.bbc.com/news/world-asia-india-37481054

Bourdieu, P. (1973). *Cultural reproduction and social reproduction*. In R. Brown (Ed.), *Knowledge, Education, and cultural change* (pp. 71–112).). *London*: *Tavistock*.

Brontë, E. (1846). *Wuthering Heights*. New York: Penguin Books.

Chauhan, A. (2016). Agra sees rise in number of domestic violence cases in 2016. *The Times of India,* Retrieved from: https://timesofindia.indiatimes.com/city/agra/Agra-sees-rise-in-number-domestic-violence-cases-in-2016/articleshow/55860667.cms.

Cockerill, M. (2019). Convergence on Common Ground: MRAs, memes and transcultural contexts of digital misogyny. In D. Ging, & E. Siapera (Eds.), *Gender hate online: Understanding the new anti-feminism* (pp. 87–110)). London, UK: Palgrave.

Coffey, A. (1999). *The ethnographic self: Fieldwork and the representation of identity*. Thousand Oaks, CA: Sage.

Dare, A. M., & Fletcher, C. V. (2019). A bird's eye view of the Malheur wildlife refuge occupation: Nonhuman agency and entangled species. *Environmental Communication*, *13*(3), 412–423.

Derrida, J. (1988). *Telepathy. Oxford Literary Review*, *10*(1), 3–41.

Deleuze, G., & Guattari, F. (1980). *A thousand plateaus: Capitalism and schizophrenia*. London: Continuum.

Foucault, M. (1978). *The history of sexuality: An introduction (volume 1)*. New York: Vintage Books.

Geertz, C. (1973). *The interpretations of cultures*. New York: Basic Books.

Goffman, E. (1963). *Stigma: Notes on the management of spoiled identity*. Englewood Cliffs, NJ: Prentice-Hall.

Graeber, D. (2013). Culture as creative refusal. *Cambridge Journal of Anthropology, 31(2),* 1–19.

Gregg, M., & Seigworth, G. J. (2010). An inventory of shimmers. In M. Gregg & G.J. Seigworth (Eds.), *The affect theory reader* (pp. 1–28). Durham, NC: Duke University Press.

Hall, S. (1973). Encoding/decoding. Media studies: A reader, 3, 28–38.

Hall, S. (1984). The Culture Gap, Marxism Today, (28), 18–23.

Hultman, K. & Lenz Taguchi, H. (2010). Challenging anthropocentric analysis of visual data: A relational materialist methodological approach to educational research. *International Journal of Qualitative Studies in Education*, 23(5), 525–542.

Jackson, P. W. (1968). *Life in classrooms*. New York, NY: Holt, Rinehart & Winston.

Larson, J. (1996). Seasons of love [recorded by the cast of rent]. On *rent* [show tune]. New York: HarperCollins.

McGranahan, C. (2016). Theorizing refusal: An introduction. *Cultural Antrhropology*, 31(3), 319–325.

Nespor, J. (1997). *Tangled up in school: Politics, space, bodies and signs in the educational process*. Mahwah, NJ: Lawrence Erlbaum Associates Inc.

Ortner, S. B. (2006). *Anthropology and social theory: Culture, power, and the acting subject*. Durham, NC: Duke University Press.

Page, R. (2000). The turn inward in qualitative research. In B. M. Brizuela, J. P. Stewart, R. G. Carrillo, & J. G. Berger (Eds.), *Acts of inquiry in qualitative research* (pp. 3–16). Cambridge: MA: Harvard Educational Review.

Parker, A., & Jensen, C. (2013). Further possible physiological connectedness between identical twins: The London study. *Explore*, *9*(1), 26–31.

Puar, J. (2007). *Terrorist assemblages: Homonationalism in queer times*. Durham, NC: Duke University Press.

Puar, J. (2013). 'I would rather be a cyborg than a goddess': Intersectionality, assemblage, and affective politics. *Meritum*, *8(2)*, 371–390.

Rahder, M. (2020). *An ecology of knowledges: Fear, love, and technoscience in Guatemalan Forest conservation*. Durham, NC: Duke University Press.

Rosiek, J. L. (2018). Agential realism and educational ethnography. In D. Beach, C. Bagley, & S. Marques da Silva (Eds.), *The Wiley handbook of ethnography of education* (pp. 403–423). Hoboken, NJ: John Wiley & Sons, Inc.

Sedgwick, E. K. (1993). The epistemology of the closet. In H. Abelove, M. A. Barale, & D. M. Halperin (Eds.), *The lesbian and gay studies reader* (pp. 29–62)). New York: Routledge.

Scott, J. C. (1990). *Domination and the arts of resistance: Hidden transcripts*. New Haven, CT: Yale University Press.

Simpson, A. (2014). *Mohawk interruptus: Political life across the borders of settler states*. Durham, NC: Duke University Press.

Trnka, R., & Lorencova, R. (2016). *Quantum anthropology: Man, cultures, and groups in a quantum perspective*. Prague: Charles University Karolinum Press.

Tsing, A. L. (2005). *Friction: An ethnography on global connection*. Princeton, NJ: Princeton University Press.

Varenne, H., & McDermott, R. (1998). *Successful failure: The school America builds*. Boulder, CO: Westview.

Vedral, V. (2011). Living in a quantum world. *Scientific American*, *304*(6), 38–43.

Visweswaran, K. (2010). *Un/common cultures: Racism and the rearticulation of cultural difference*. Durham, NC: Duke University Press.

Weheliye, A. G. (2014). *Habeas viscus: Racializing assemblages, biopolitics, and black feminist theories of the human*. Durham, NC: Duke University Press.

Wolcott, H. F. (1994). *Transforming qualitative data*. Thousand Oaks, CA: Sage Publications.

Wozolek, B. (2020). Hidden curriculum of violence: Affect, power, and policing the body. *Educational Studies, 56(3)*, 269–285.

2

GORDIAN KNOTS

Grasping At The Impossibly Intertwined

ANKITA'S NARRATIVE: *At first, he said it was an accident, you know? He said he had been drinking too much and, so, when he forced himself on me, he said he didn't mean to hurt me. It's just that I looked so beautiful that he couldn't control himself. What I remember about that first time isn't what happened, it's what happened after. He was so kind. He brought me flowers and even helped with cleaning up the kitchen after. Sometimes I think I can still smell those flowers. They were roses that seemed to be tinged by the smell of jasmine from the market. We hadn't been married for long. In fact, we had just moved from his mother's house into our own place. It became more frequent after that. I felt so alone. In those moments. In my own body. Even in my own family. How could I talk about it? It's not rape if it's between a husband and wife. I know that. My mother once told me that about some girl who went on publicly about her husband who did these things. Anyway, I took it as long as I could because he's a good man. And, let's be honest, I need to be a good wife. My husband, he works hard and makes money. He also comes from a good family. I can't disgrace my family over this. So I'm just here, doing my best. Thinking about if I should go back and just end this whole court process. It's messy. I don't want this mess.*

TAARA'S NARRATIVE: *Sometimes, I just knew it was going to happen. It was as if I woke up and "tun", it just hit me. Sometimes I didn't know that I knew until later. I would reflect on the day and remember feeling like I knew. This one time I remember hiding in the bathroom. He was so mad at me for… I don't even remember what. I ran in and locked the door. Even as I felt the bolt snap into place, I knew it wouldn't be enough. He broke down that door in minutes and then I got it worse. I left because once I saw my daughter watching. She wasn't upset. She just stood there, watching, like she was figuring something out. What all do you suppose she was figuring? I'm still afraid to know.*

RINA'S NARRATIVE: *Sometimes violence is so small. It's hard to know it's there if you don't know what you're looking at or looking for. It started with my cooking. He would always tell me to do it this way or that way, to be more like his mother. Then it was the way I dressed, or looked, or wore my hair. Pretty soon I began to feel like there was nothing I could do right. I couldn't pick out the right food at the market. I couldn't instruct the maid how to clean properly. It was always something. And I always work hard to do my best, to be a good wife like my mother to my father. Then he started yelling. Always yelling. The neighbors would hear it was so loud, although they never said anything. That's the way things are around here—it's never anyone's business. Sometimes these things come out in conversations over tea, with other women, but it's like they hear you but have no idea how to help. They hear everything. They never help. He's never hit me. He's never hurt me. But he does. He does hurt me all the time with the yelling and telling me how I'm no good. How I'm worthless. I just wish I had some value. To someone.*

Several times in my adulthood I have run across this question on social media: "When did you first encounter racism?" The comments are generally filled with answers that vary from, "When I was five…" to, "It wasn't until I was a young adult that I understood how racism impacts communities of color". Generally, the comments show children of color experiencing their entree to racism painfully early in life while white respondents come to an understanding much later. Scholars such as Feagin and Ausdale (2001), Love (2019), and Grace and Nelson (2019) have argued similar points regarding learning about race and racism, giving accounts of children as young as three describing racist experiences. For me, the first experience with racism[1]. I remember was in kindergarten, when I was kicked by a classmate and told that "dirty-skinned kids go to the back of the line".

This chapter focuses on violence that is related to race, gender, and sexual orientation. The question could just as easily be, "When did you first notice gender discrimination or homo/transphobia?" Speaking to the opening narratives, the question is often, "When did you first experience violence in your relationship?" While experiences are far from universal, marginalized people and groups are disproportionately affected by cis-heteropatriarchy (Johnson, 2016, 2016; Love, 2019).

The violence a person experiences can be thought of on a spectrum between direct and indirect aggression; both effect and affect the body. Direct violence—personally encountered aggression like being sexually assaulted by one's partner—has been shown to impact the body in deep and perhaps permanent ways. These impacts include suppressed immune system function, unrecognized physical consequences of violence, post-traumatic stress disorder, and anxiety (Bohn & Holz, 1996; Daphna-Tekoah, 2019). Direct violence in the form of verbal aggression can change a person's blood pressure and predispose them to a multitude of stress-related illnesses (Anand et al., 2019; Bourassa et al., 2004). Similarly, indirect violence has pushed researchers to call for an expansion of the term "epigenetics"—changes in the physical structure of a person's DNA—to include the idea that trauma can leave a chemical mark on a person's genes that is passed down to subsequent generations (Yehuda & Bierer, 2009). Trauma therefore exists like an echo chamber that reverberates on the body, across time and space.

These recursions of violence do not solely impact human bodies. Sharpe (2016) argues that the ocean still swells with the violence of slave ships traversing its waters. She builds this argument by looking at the time it takes for a substance to enter and then leave the ocean, known as "residence time"; sodium has a residence time of 260 million years. The human body, Sharpe writes, "is salty [and] what happens to energy produced in the waters? It continues cycling like atoms in residence time...Black people exist in the residence time of the wake, a time in which 'everything is now. It is all now'" (p. 40). This argument is literal, not metaphorical. Salt is shared and recursive. Violence is literally in the salt, in the water, in our bodies. But it is also metaphorically true; this is where things get complicated. All conditions, like saltiness and violence, are always at once metaphorical and material.

Broadening Sharpe's argument, the ocean's body has likely been impacted by countless traumas. For example, it is impacted by the violence of pollution. Roughly eight million metric tons of plastic waste are deposited annually, breaking down and found in both water and sea life (Ostle et al., 2019). Much like the human body, plastics have a long residence time. Microplastics will continue to write and rewrite the health of the ocean in immediate and generational ways. I note these effects not to equate pollution with the enslavement of Africans as chattel, but rather to draw attention to the diversity of trauma the ocean body experiences.

Returning to human bodies, women in India rank among the highest in the world for suicide (Armstrong & Vijayakumar, 2018).[2] Local papers have noted how the national suicide rate impacts coastal communities as lifeguards struggle to try and save women who would otherwise die by suicide in the ocean (Das, 2018). Oceans cover 71% of Earth's surface. From the atrocities of enslaving people to the trauma that provokes self-harm to the disregard for ocean life with increasing pollution, the amount of trauma churning in the sea is incredible and incalculable.

Violence is therefore not just something shared affectively. It is also found in the molecules of bodies or, in this case, in the body of the ocean. More generally, violence is in the ocean, in the grass, or in one's exhale in the moments before, during, after, and between violent encounters. Because violence is both direct and indirect, as trauma becomes a part of bodies like the ocean, time collapses on itself. The trauma of today is appended to yesterday's violence. One way of understanding this is through the block universe theory in which time is relative and the past, present, and future of every time and place exist together. Tomorrow's aggressions have, in a way, already happened.

Take Taara's narrative from the chapter's opening, in which she anticipated violence before it hit. While one interpretation of her statement might be that she was responding to affective tensions, another might be that through previous intra-actions with her partner, violence did more than just stick (Ahmed, 2010) on her ways of being; it entangled across time and within the multiplicity of violent encounters. This entanglement extends from the experiences Taara had with her husband to Ankita's narrative of the normalization of violence within community that silences women when they want to speak up. As described in Taara's narrative, this violence can also extend to the next generation's ontoepistemology.

Beyond Einstein, who argued that time passes differently for people moving relative to each other, the fluidity of time is a well-trodden dialogue across the globe, from Mbiti's (1969) work on Shasha and Zamani to the Jain philosophy of kāla, which distinguishes between time as substance and time that is measured. Time, Champat Rai Jain (1915) wrote, can be:

> conceived as parts of the substances that revolve around them, is obvious from the fact that they are necessary for the continuance of all other substances, including souls and atoms of matter which are simple ultimate units, and cannot be imagined as carrying a pin each to revolve upon. Time must, therefore, be considered as a separate substance which assists other substances and things in their movements of continuity.
>
> (*p. 528*)

Fluid conceptualizations of time have also been discussed by philosophers like Skow (2018) and physicists like Austin (2017), who recognize the fluid nature of a subject that is too often imagined to be concrete.

Affect like trauma can be understood within a similar framework. Trauma has often been thought of as resolving or taking up less emotional space with time. However, when a familiar smell or sight evokes the memory of trauma, a survivor can be pulled across time, recalling the pain of the event (Brewin, 2015). In addition to survivors' narratives, Sharpe (2016) asks what happens when those who have not directly experienced trauma look at violence across time. Similar to scholar like Hartman (1997) and Moten's (2003), whose interrelated work attends closely to the nuances of witnessing, Sharpe argues for a form of witnessing that does not retraumatize or subjugate the survivor through the gaze of the witness but, rather, burdens the witness ethically. Such observation demands that a witness imagine other outcomes, pasts, presents, and futures (Douglass, 1855; Moten, 2003; Wiesel, 2002). Even in witnessing, time collapses, and despite being distinguishable and particular, present and past places, spaces, and bodies are enmeshed (McKittrick, 2006, 2015, 2015; Sharpe, 2016).

This entwined nature of bodies has similarly been discussed by scholars like Woodson (1933), Puar (2007), and Sabzalian (2019) as they examine the normalization of sociocultural ideas and ideals through relationships. For example, to Woodson's (1933) point and borrowing from Helfenbein (2010), cultural norms, values, and actions *speak* and *leak* between the classroom and communities. While I turn in greater depth to questions of schooling in Chapter 4, it is worth noting briefly here how the circulation of sociocultural ideas and ideals begin and end in the classroom. From the everyday violence against brown and Black bodies in the United States (Love, 2019) to the pervasive proliferation of sexual violence across the globe (Phipps et al., 2018), schools are often important spaces in the webs of significance that connect cultures and people.

One way to conceptualize the relationships between time and intra-action is through assemblages. Several inroads in assemblage theory have been significant to posthuman and new materialist constructions of relationships (e.g., DeLanda, 2016; Deleuze & Guattari, 1980; Puar, 2007; Weheliye, 2014), along with other conceptualizations across the social sciences that get at similar ideas of the nested layers of complex and emergent relationships (e.g., Agar, 2004; Geertz, 1973; Gershon, 2017; Massey, 2005). Because of the overdetermined use of the term through its multiple iterations, one can argue that the term "assemblage" has become an empty signifier. As Deleuze and Guattari (1980) theorize, assemblages can be significant to reconsidering relationships and the bodies that form and inform matter. However, as Puar (2012) argues, what an assemblage *is* often eclipses and what an assemblage *does* in these analyses. Overdetermination, Puar argues, happens when an idea is confined to describing the parts of an assemblage rather than theorizing the possibilities and consequences of the assemblage's movement. Similar to Puar (2012), rather than focusing on what an assemblage is, in this chapter I attend to what an assemblage does.

Contemporary dialogues across the sciences focus on what is produced through assemblages. By assemblage, I refer to the messy and entangled intermingling of bodies that is productive in that it gives "expression to realities, thoughts, bodies, affects, spaces, actions, ideas, and so on" (Weheliye, 2014, p. 47). Within a Deleuzian definition, assemblages are also characterized as not being "uniform either in nature or origin [while] actively linking parts together by establishing relations between them" (DeLanda, 2016, p. 5). Much like identity, assemblages are as much about in-group as about between-group intra-actions.

While assemblages can be theorized through the human body and its intra-actions, (e.g., Puar, 2007; Weheliye, 2014), assemblages also exist between nonhuman entities. For example, scientists have recently discovered that a spider's web is not just a tool through which the arachnid senses its prey, it is a form of extended cognition. In other words, spiders don't just use their webs to sense, they use them to conceptualize (Robson, 2020). Scientists are starting to use this theory of extended cognition to understand how plants "think" (Parise et al., 2020) and to reenvision mathematical understanding as an embodied way of knowing (Walkington et al., 2019). To be clear, the examples of plants and spider webs are not just cases of anthropomorphization. While the human terms "think" and "conceptualize" are used here, it is my impression that, given time, we will learn that what forests (Kohn, 2013) and spiders do is far more interesting than what we understand as "thinking". One could therefore think about an assemblage of cognition, or how affective relationships (re)produce relationships that circulate between bodies. Returning to social media, one only needs to think about the assemblages that exist among social media users to better understand how relationships are reproduced through webs of intra-action, as young people in particular use these media webs to think about their own personhood and positionality on local and less-local scales.

Assemblage theory produces another way to examine the intraconnected nature of bodies, as scholars like Delanda (2016), Deleuze and Guattari (1980), Puar (2007), Weheliye (2014), and others have noted. This is not to forget the significant contributions of anthropologists and sociologists who have also discussed complicated webs of relations (Geertz, 1973; Kawada, 2019; Massey, 2005; Rosaldo, 1993; Tsing, 2005). However, as I noted in the introduction, my use of assemblages pays attention to both sociocultural relationships and to the way we entangle with others on molecular levels, both theoretically and materially. If, as Barad (2007) argues, intra-actions are productive, then it is worth explicating how we entangle across spaces of being, knowing, and doing. Intra-actions always produce agency and shift ways of being, knowing, and doing to some degree. Just as the ocean still holds the literal and metaphorical essence of human violence, assemblages impact bodies across sociocultural and subatomic levels in literal and metaphoric ways.

Assemblages are also ubiquitous. Much like Foucault's (1976) argument about power, there is no outside of assemblages. They are therefore as much about power and agency as about anything else. The difficulty with this pervasiveness is that thinking about one structure can often lead to falsely bracketing others. For example, although the previous chapter focused on how hope can spread, the underlying idea was still about how hope and violence are interconnected through intra-actions. In attending to violence, such an approach might mean focusing on aggression rather than an assemblage of hope, but that does not make the two any less entwined. Further, while expounding on the contents and products of assemblages can be useful for focusing on a particular affect, such explanations can also flatten experiences and sociopolitical structures like race, multiple forms and expressions of gender, sexual orientation, and the like. Weheliye (2014) argues that texts like *A Thousand Plateaus* often leave questions of race to the margins, such that an assemblage theory centered solely on a Deleuzian notion of the rhizome can potentially (if incidentally) be oppressive in its lack of anticolonial complexity. Even within assemblage theory, then, there are assemblages of violence.

As I mentioned in the previous chapter, a significant portion of my work as a scholar has focused on violence or what I have more recently come to call an *assemblage of violence*. I use assemblages of violence to attend to the messy, entangled sociocultural and political practices that exist in relation to human and nonhuman entities, practices that are constantly in a process of being sewn together and pulled apart. This is what Deleuze and Guattari (1980) call *deterritorialization* and *reterritorialization*. In more simple terms, deterritorialization and reterritorialization are akin to when students use their seating to gain social territory in a classroom: when, one day, one person sits someplace different, the classroom can have a different structure, both physically and culturally.

In this chapter, I specifically attend to how systems are infused with violence. Weheliye (2014) discusses the identity of an assemblage and the subsequent identities formed through that assemblage as bodies and beings that are in precarious relationship to each other because of the constant deterritorialization and reterritorialization central to meaning-making processes. I am similarly interested in identity, the content of an assemblage, and how that content operates as a part of an assemblage in practice.

Assemblages of violence describe the many trajectories and possibilities that multiple forms of oppression and marginalization express on and through bodies and ideas—be they intellectual, emotional, or physical.

Discussing violence as an assemblage recasts the singularity through which violent interactions are often discussed. Instead, using assemblages of violence insists that any one iteration of violence is not singular; it is always necessarily dependent on sociocultural norms, histories, and other intra-actions of violence.

Take, for example, the narratives at the opening of this chapter. On one hand, when a woman faces emotional or physical violence, it requires multiple intertwined systems of normalized misogyny to materialize as a single act. On the other, the normalized way that women in any abusive context (independent of place) are expected to adjust or disregard that single act—or successive iterations of that act that are treated as separate singularities—does not change overarching systems of oppression. The same can be said of a person who has racial or homophobic slurs yelled at them or those who are recipients of ableist forms of oppression and aggression. In other words, any one act of violence is necessarily part of systems, individuals, and ecologies, including other systems of violence and also, complicatedly, systems of peace and justice.

As aggressions occur, they do so in connection and intra-action with broader assemblages. In the chapter's opening narratives, each seemingly singular act of violence folded and refolded into assemblage, like a never-ending Mobius strip that reterritorializes each act into similar yet distinct patterns in spite of their commonalities. While each intra-action reterritorializes questions of agency, past encounters with violence are also at play. In Rina's case, acts of aggression against her were noted but not interrupted by her family and neighbors. In this way, past aggressions impacted Rina and her husband but were also shaped by the norms and values of her family and neighbors as they were woven into the assemblage as witnesses.

In addition, acts of aggression can be understood as coming from within an assemblage, in both form and function. As violence emerges from an assemblage, sometimes bubbling up slowly and sometimes erupting suddenly, the assemblage itself is temporarily deterritorialized by the movements of violent intra-actions. The assemblage is an ever-knotted thing, always in motion. As it moves, the violence that it is and that it contains emerges in unpredictable yet patterned pathways. The previous form of the assemblage is deterritorialized as movement changes the assemblage from what was into what will be. That which is reterritorialized, as scholars like Barad (2007) and Rosiek (2018) argue, is coconstituted between bodies that intra-act within the event. As agency, multiple agentic contingencies, and other aspects of power are coconstituted, the assemblage is reterritorialized as it folds back into itself.

Think for a moment about the tension felt at any event focused on social equity where people with opposing perspectives are present. In this case, I use the example of Vanitha Mathil, or the "Women's Wall", that occurred in India on January 1, 2019. This chain stretched across the Indian state of Kerala and was a symbol of people united in a protest against gender-based discrimination. Specifically, it was a protest against the ban on women between the ages of 10–50 from entering the Sabarimala Temple because they could be menstruating, something considered culturally "unclean" by many devout Hindus across the country (Deo & Ghattargi, 2005). In 2018, the Supreme Court of India lifted the ban, but the temple upheld its original position, going against the court's ruling. The human wall in protest of the temple's ongoing sexist position extended 620 kilometers (390 miles) from Kasargod to Thiruvanathapuram. The event was covered across the world, from NPR to the BBC, from the *Times of India* to *Time* magazine. The day after the protest, two women were escorted by police to worship at the temple and then taken to an undisclosed location for their safety. The temple was "purified" after they left, and one woman was disowned by her family for worshiping at the temple (Berger, 2019). Lead by conservative men's rights groups, violent counter protests erupted, seriously injuring several people.

While global news coverage ebbed after Vanitha Mathil, the battle did not end. In November 2019, the temple opened for its annual pilgrimage and the government in Kerala declared that the shrine was not a place for activism. Over a hundred police officers were deployed to the area but, as stated by the government, they were not there to guarantee protection for women trying to worship at the shrine or participate in pilgrimage activities. Instead, they were there to keep the peace during a time of worship. As of 2020, the temple remains a site of contention, with the Supreme Court currently accepting petitions to overturn their initial ruling. News outlets in India have reported that several women have tried to enact their right to worship at the temple but have not made it inside.

Vanitha Mathil was an event of solidarity, one that women described as the "dawn of a new era of feminism in India" (Thiagarajan, 2019, p. 1). The other side of that coin is that while the wall was successful in raising awareness and creating agentic spaces for women and girls, women in the state still encounter waves of violence, including from police and a government who are unwilling to guarantee their safety (India Today, 2019) and from increased attacks on women attempting to enter the temple or who simply live in the area (Berger, 2019).

The aggression surrounding Vanitha Mathil can therefore be understood as coming in waves of varying intensities and as a constant unfolding and refolding of assemblages of violence that happen through ongoing events. As the wall was being organized in the months leading up to the protest, violence swelled. Afterward, violence erupted more intensely as men rioted to show their displeasure with women entering the temple. However, the events surrounding Vanitha Mathil did not begin or end with/in the event. Women in India have faced centuries of discrimination centered on being banished during menstruation (Du Toit, 1988). In the state of Maharashtra, for example, women are still relegated to *gaokors,* or menstrual huts. These poorly constructed dwellings often leak during monsoon, are sometimes invaded by wild animals, and can become so swelteringly hot that women have died of heat stroke while they wait for their period to end. There are no kitchens in these spaces, as menstruating women are not permitted to cook. This leaves women are at the mercy of their families to bring them food during this time.

While one outside of India could easily dismiss this as a local issue, the stigma around menstruation is global. In the United States, for example, menstrual products are taxed, with some states just beginning to consider eliminating the tax. Importantly, over half of women across the country describe "menstrual shaming" as a frequent experience (Fahs, 2016). This extends to schools, which frequently require girls who are just getting a handle on how to negotiate their periods to wear khaki pants, increasing the chance that breakthrough bleeding will be noticed and criticized by their peers. In October 2019, female Japanese employees at a department store were encouraged to wear badges denoting that they were on their periods in an attempt to "improve the working environment" by sharing the information. The result was that many women faced harassment and were "reluctant" to wear the badges (BBC News, 2019). In Britain, the government created an entire taskforce in 2019 to fight the stigma of menstruation globally and locally (Rivera Antara, 2019); this stigma is experienced by more than a third of women in the UK (ActionAid, 2018). The violence that women in Kerala experience around their periods is not just local to the state or the country, it is an international norm. Within an assemblage of violence, contexts are connected even when not necessarily in direct relation. Just as the narratives above stand on their own, they are also connected through assemblages of violence across space, place, and time.

An assemblage is how things—affects and actions—manifest. This is because assemblages *are* and *do* simultaneously (Deleuze & Guattari, 1980; DeLanda, 2016; Puar, 2012; Weheliye, 2014). Violence is both an assemblage and a body, meaning that the assemblage itself can be understood as relational and agential in ways that can propagate violence. In sum, much in the way that culture or racism was discussed in terms of bodies above, assemblages are also agential (Bennett, 2010). Bennett (2010), for example, argues that objects are not passive but rather dynamic and vibrant in their interrelationships and trajectories. Assemblages of violence are similarly dynamic in their trajectories.

Returning to the protests in India, batons wielded by officers carried affective messages. During and shortly after Vanitha Mathil, the police were agents of protection. However, women's ability to act was soon limited by systems of power in the form of agentic contingencies that were, in this case, set by affective tensions. Specifically, the police carried with them the affect of violence against activists, especially as protections were no longer guaranteed. These messages were contextualized not only in the moment but across time, in relation to how and where women and others who have defied police orders have been struck down by police. Affect travels and, as Bennett (2010) explains, affective bodies are associative in that they contain a corresponding and inseparable capacity to affect and to be affected. Objects like batons never act in isolation. These relations between things form and inform assemblages, which, as the sum of their parts, are agential. Similar to Barad's discussion, because agency cannot be localized, it becomes distributed in coconstitutive ways across those entangled in an assemblage.

Finally, assemblages are unpredictable in their fluidity, yet patterns do emerge from within an assemblage. Power, for example, is and will always be ubiquitous, though as Ortner (2006) argues, our ability to act and react within systems of power can vary. Because agency is enacted with intention (Barad, 2007; Gershon, 2018; Ortner, 2006) and assemblages are agentic, they engender and maintain certain norms and values. For example, in the narratives above or in the example of Vanitha Mathil, the violence inherent in patriarchy remains constant and consistent. However, because quantum entanglements occur within assemblages, power is often nested with spooky actions that cannot always be predicted. Yet because not all entanglements within an assemblage are quantum entanglements, the complex body of the assemblage remains unpredictable in how it will be deterritorialized and reterritorialized within its intra-actions. In other words, while one might be able to conceptualize the boundaries and borders of an entanglement, when entanglement happens at the quantum level, those knotted within an assemblage as part of the quantum entanglement might not be aware of the close relationship that feels as if it came out of nowhere.

This is important because within an assemblage of violence, sometimes hope is the only thing you can hold on to. To be clear, I do not present hope in some kind of duality with violence. Rather, I mean that things like care or hope are often what people cling to in the face of violence. One might theorize that this is because hope exists as its own assemblage in which violence is subsumed. Or perhaps one might frame a single giant assemblage of affects where emotions are a spectrum, in which one sometimes consumes the other in moments of happiness. I am not discounting these possibilities; however, my focus here is on violence. Toward that end, while an assemblage of hope is always necessarily connected to violence in implicit and explicit ways, the ability to reterritorialize agentic structures within their instability is important to allowing affects like hope, joy, and love to be central within lived experiences.

In the following section, this chapter turns to three narratives from youth at New Horizons. These narratives further show how assemblages of violence function. These three narratives were chosen specifically because, while they express different experiences across sexualities and gender identities, they are similar in their illustration of affects like shame. Following these narratives, I explain how shame is central to the movements within an assemblage of violence.

Violent Assemblages Matter(ing) in Middle School

TANISHA: There's this one teacher. He's a little, you know, touchy touchy. It's sooo creepy. And then he follows us to the bathroom. He doesn't come in. He just stands there and when we come out he puts his arm around us and says things like, "You need to hurry up, okay? You can't take so much time in there". Look, I don't know what I am. Am I a lesbian? Maybe. I haven't had a girlfriend. Am I straight? Maybe. I haven't had a boyfriend. But what I do know is that I don't like it when Mr. G does that. It makes my hair all standing up like, "Woah, no thank you" and it makes me feel bad about myself. The issue is that I can't really say anything. My mom, she thinks this school is the best place. I'm not trying to make this school a bad place. It's good, you know? I can get into a good high school, or at least an okay high school. And it's not like this kind of thing is new, either. Men be out there doing stuff to women all the time. And, also, he didn't do nothing. Just put his arm around me. And not just me. Lots of girls. Well, Black girls. Sometimes Hispanics but never the whites. What else can you do? It's just normal.[3]

DANIEL: The bathrooms are down again. We can't pee from lunch until the end of the day. They are locked because people be messin' around in there. I just feel sad about it because I used to have time with a boyfriend in the bathroom. It's silly. Don't you be giving me the business about this but…we would go there when we needed to cry. We'd just hold each other and, well, cry. Sometimes you just need to have a good cry 'cause things aren't always right. People be tellin' you how to be and who to be. And me? I just wanna be me. Like being me, [*softly*:] *gay,* isn't a man so it's not right. And that makes me want to cry. Okay, so, maybe he's not a "boyfriend". He's a cousin though. If you gay, or whatever, you have to find people who are your family. People can love *you* for *you.* The people who let you hope to be whatever you want to be. So when I had to, I'd just text him and have a cry. That's all. Nothin' bad. We didn't even mess with the soap or mess up the paper towels.

WOZOLEK: You said, "Hope to be". Can you tell me more about that?

DANIEL: Look, I'm Black and I like boys, and sometimes girls, maybe...I don't know right now. But, in my life, there's not hope. And there's not a lot that makes me hope. School should do that, right? This is supposed to be a good school. But I'm always in trouble, so what can I hope for? My white-ass teacher once told me family can do that. But they don't love me while I'm gay. But being near [my boyfriend] does. You gotta choose the people who lets you hope. Not everyone. Not every place you're in does that.

EMMA: I don't go to the bathroom. I hold it. Every single day. Once I had one of those bladder infections. The doctor at the clinic said it was probably because I wasn't drinking water or because I was not going to the bathroom all day. But why would I go in there? I don't really belong in the girls' bathroom. And if I went in the boys, I think someone would get me for sure. I'm a white kid in a mostly Black school. I think that makes it worse. Like I don't already stick out because of who I am, I have to stick out more because of, well, who I am. But, oh, when I get home, you know that's the first place I go—to the bathroom. This school is nice, you know. The teachers are nice. The counselor is nice. But you know the one rule of the school, right? Don't embarrass the principal. And that's basically me. I'm an embarrassment. That's what my stepdad said anyway. So, that's me, I guess. The kid who doesn't pee.

Capitals of Shame Within the Assemblage

Affects are shuttling intensities (Seigworth & Gregg, 2010) that entangle, moving in and through an assemblage. The narratives throughout this chapter offer distinct descriptions of violence and its interrelated affects that overlap in important ways. Shame, for example, is a common thread among experiences—from women who were ashamed to speak out with family or neighbors to students who felt shame in using a restroom to cry, to urinate, or because of feeling surveilled. Although there are several intensities that exist in an assemblage of violence—rage, fear, terror, and anger, as well as hope, joy, and pride—here I focus on shame. In contrast to the way that guilt is associated with *doing* something, shame, as Sedgwick (1993) argues, is "peculiarly contagious and peculiarly individuating...a movement toward uncontrollable relationality...toward the implication that one *is something*" (p. 37). In other words, shame is far more insidious than guilt because it is meant to be a reflection on who a person is rather than on what they do.

There are many ways to understand how shame travels between people. Psychologists have placed shame in relation to rage (e.g., Kohut, 1972) or as part of a cycle of shaming (e.g., Nathanson, 1987, 1987). Anthropologists have used shame to talk about questions of gender and culture (e.g., Lindisfarne, 1998) and sociologists have discussed shame as it intersects with the sociopolitical topic of hair (e.g., Synnott, 1987). Finally, queer theorists have had several dialogues regarding queerness as it is linked to shame and shaming (e.g., Cohen, 1997; Halperin & Traub, 2009, 2009; Munt, 2017; Sedgwick, 1993; Stockton, 2006).

Following Munt's (2017) discussion of the queer attachments of shame, shame is an emotion that can occur in both intense and quiet moments. Shame can linger on and through the body, but also exists between bodies. Although shame is often perceived as negative, something to be expelled from the psyche, degrees of shame can be understood as productive. Munt argues that "we require a degree of shame to perceive moral transgression" (p. 4). The idea of a "moral transgression" is complicated because sociocultural values might not always translate to equity. Colonization, eugenics, enslavement, and genocide might not be understood as morally corrupt to those in power, and therefore shaming might not be impactful to those who believe they are on the "right" side of heinous acts of violence. The opposite side of the coin is that those fighting against such atrocities are often shamed and marginalized for resisting or refusing acts of oppression. The act of shaming as a moral imperative is therefore slippery in that it is a sociopolitical and cultural construction.

While many social theorists and medical professional conceptualize shame outside of assemblages, these ways of positioning shame tend to fall short. This is because shame, like violence, can be theorized through seemingly singular events. Here, I conceptualize shame in two key ways: First, I position shame within assemblages of violence. Placed within an assemblage, the act of shaming is not isolated. Through this perspective, that one act serves as a manifestation of the broader assemblage that is always necessarily connected to past, present, and future iterations of aggression through shaming.

Second, as I have theorized elsewhere (Wozolek, 2019), it is also useful to think about shame as a form of capital, or as *capitals of shame*. This is not unlike the characterizations of shame above, in that capitals of shame allow one person to shame another for their ways of being and knowing in order to engender and maintain systems of power. Many dialogues on capital (Bourdieu & Passeron, 1977) have been discussed across queer spaces (e.g., Britzman, 1995), critical race dialogues (e.g., Yosso, 2005), critical feminist conversations (e.g., Valenzuela, 1999), and across forms of qualitative research (e.g., Pasco, 2003). The frame of capitals of shame is aligned with dialogues on cultural capital (Bourdieu & Passeron, 1977), which describe the societal assets of a person that promote social mobility. For example, the girls at New Horizons often discussed the ability to sing as something that gained them social mobility among other students; there were several sessions where students spontaneously broke into song, looking around to see how their voices were perceived. While the talent to sing is not necessarily tied to financial capital, the ability to take singing lessons was imbricated in how some girls' abilities were nurtured over others'.

Capitals of shame are how those with privilege and power negatively employ sociocultural norms and values, at once cultivating criteria for shaming and then using those criteria as a means to declare people or groups publicly noticeable in their incompetence or impotence across all manner of interactions, possibilities, or ways of being, knowing, and doing. However, a capital of shame is not a form of power that one wishes to gain, as opposed to cultural or financial capital for example, but is instead the ability to use one's available wiggle room to shame others (Wozolek, 2019). As with all forms of capital, those with less power can still use their agency to shame the powerful, either individually or as a group. Vanitha Mathil offers but one example of how one movement can bring international shame to those in power. This is especially true in an era where social media can raise awareness around the globe.

However, those with capital are much more likely to shame others and to understand those shaming actions as normal and socioculturally appropriate. In other words, capital still functions like capital, and shame still functions like shame. For example, in the Indian context, women's shaming is carried out by people regardless of their gender. Mothers or female neighbors were as likely to shame women for not being "good wives" as husbands, fathers, and other male community members. At the middle school, Emma discussed how her ability to use the restroom, an act that should be a fundamental right, was so often shamed that she abstained from using the toilet. Who she was in the world was so shamed that it impacted her bladder health, yet such shaming continued consistently and constantly during the school day. Such forms of shaming are so common for trans and gender-nonconforming youth that studies have shown increased rates of student suicide in places that lack gender inclusive restrooms or where students are denied the ability to use a restroom that is aligned with their gender identity (Sutton, 2016).

The ability to gain capital from shaming often functions like an echo chamber of violence within the margins of intersectional identities. An echo chamber is an enclosed space where sounds reverberate. One sound continually ricochets off the walls, both scattered and absorbed by the surfaces as it travels. What one hears in an echo chamber is the complicated series of sounds that come from echoes reverberating off many surfaces, combining together in the ear to render it sensible for the listener. Add additional sounds in an echo chamber and the noise becomes chaotic. Imagine for a moment that shame reverberates like sound, with a socially isolated person or group in the echo chamber. In this case, the people in this group are lesbian transwomen of color. They are shamed for their gender identity, expression, sexual orientation, race, and possibly for body image within white standards of femininity. The "sound" of shame reverberates tumultuously, and this shame is both absorbed by the people experiencing it and bounces off of one marginalized person to land on another.

The scholarship on how Black girls are sexualized at a young age (e.g., Morris, 2016) is another way to examine how shame operates like sound. Tanisha's words underscore the echo chamber of shame (being a Black girl) as it has been normalized and, in this case, is met with aggressive actions. At the intersection of queerness and race, shaming by the white cis-heteropatriarchy has become so normalized that, while Tanisha cannot necessarily name the relationship, she feels the affects and effects of sociocultural shaming in broad and local circulation.

Shame therefore functions within assemblages of violence as a form of aggression that impacts people across ontoepistemologies. One of the reasons that shame is so insidious within assemblages of violence is that it is often positioned as binary with pride (Derrida, 2008; Halberstam, 2009; Halperin & Traub, 2009, 2009; Munt, 2017). Much like the way that hope is often entangled with violence, pride is similarly connected to shame and shaming. Rather than understanding these affects as binaries, I propose that they are knots of possibilities (Berlant, 2011) that are territorialized and deterritorialized through their intra-actions.

Even in the #MeToo era, despite the rise of global pride in women's ability to voice their rights to their own body and safety, this pride is stifled when stories of sexual assault are shamed back to the margins. Capitals of shame not only allow the person who is shaming another to garner and maintain power, but also limit the agency of the person who receives shame. If Barad's framing of intra-actions is correct, this process of gaining power through shame is part of the coconstitutive nature of the intra-action found in shaming and experiencing shame. The ability to shame another therefore acts as an agentic contingency over the shamed's ability to move through an assemblage of violence.

As shame travels through an assemblage, it can be thought of like a magnet, both pushing and pulling on other affects. While this movement does not always exist in a polar binary, magnets are a helpful metaphor because they produce electromagnetic force through the motion of their charges. Assemblages of violence are also productive because of movement caused by affects such as shame, which produce force and friction. Just as shame moves and is moved between bodies, assemblages of violence move bodies as much as they are moved by bodies. As Weheliye (2014) notes, violence—and the assemblages entangled with and driven by it—is borne from a sociopolitical project to maintain difference. In the lives of the participants in these studies, while they frequently worked to have relief from the violence in which they were entangled, they often found that there was no "out", only the ability to have, keep, and fight for agency within. In other words, while extracting themselves from ongoing conditions and iterations of violence was difficult, they still had agency to move within those conditions. This is why an assemblage of violence is a helpful construct: it is a way to conceptualize how those who work so hard to escape violence remain ensnarled in it.

Shame is but one example of a driving force within an assemblage of violence. It is also an example of what an assemblage of violence can do. Assemblages of violence act to normalize aggressions that fall on and between bodies. This is because of the entangled nature that human bodies have with violence. What violent assemblages *do* is twofold. First, assemblages of violence are instructive; this is a point to which I turn in greater detail in Chapter 4. Second, akin to Foucauldian (1978) arguments about biopower—sociopolitical power used to assert control over actors' lives and bodies, fostering life or to denying it to the point of death—assemblages of violence are instrumental in asserting and normalizing control. The women in India, for example, discussed pressure to stay with their husbands while experiencing few sex-positive relationships and intense fears for their physical and emotional safety. Similar to the students at New Horizons, they articulated a feeling of constant surveillance. While for the students this was a question of being surveilled at the restrooms, for the women it was being watched as they carried out their duties of cooking, cleaning, or being a "good wife".

These feelings of being surveilled and controlled at home are mirrored in national politics. As but one example, the Indian government continues to exercise control over women and girls through events like those at Sabarimala Temple. This normalization of control and aggression against women who resist or refuse the norm has a strong trickle-down effect to the daily experiences of women in their homes. At the middle school, the intrinsic and extrinsic forms of control within systems of schooling also leak into intra-actions with peers, teachers, and administrators. Assemblages of violence are entwined with local and less-local forms of control in ways that encourage and, in many cases, limit people's ability to resist or refuse violence that impacts their lives. For some, there comes a point where conforming to violence is easier than resisting it.

In addition to Foucauldian biopower, assemblages of violence are theoretically connected to Mbembe's (2019) use of necropolitics and Puar's (2007) discussion of queer necropolitics. Briefly stated, necropolitics are the many ways that social and political powers are used to dictate how some people may live and how others must die. Mbembe is clear that while Foucault uses biopower to discuss sociopolitical control over citizens, necropolitics specifically addresses political violence in the form of the right of those in power to enslave, maim, or kill others. In other words, sociopolitical systems of control are linked to the subjugation of life to the power of death. Mbembe argues that necropolitics makes "the murder of the enemy its primary and absolute objective to engage subjects in a unique form of social existence in which vast populations are subjected to conditions of life conferring upon them the status of *living dead*" (p. 39, author's emphasis). Marked by colonial encounters, necropolitics explores how some populations are effectively marked for death by others.

Much like in the narratives of the students at New Horizons, Puar (2007) uses Mbembe's work on necropolitics to engage the ways that "queer" is constructed across narratives that disrupt normalized ideas and ideals, existing in tension with normativity. Queer necropolitics takes up similar questions to those raised by Mbembe, but attend specifically to "queering the necropolitical [while interrogating] claims to queerness in the face of death, both spectacular and banal" (Haritaworn et al., 2014, 2014, p. 2). Attending to tensions within Judith Butler's work on how queers are left to die, Puar asks how and which queers reproduce and are folded into life and, subsequently, which queers give their lives. Queer necropolitics evoke the "production of disavowed subjectivities, socialities, kinning, intimacy and desire while bringing into sharp relief the consolidating alignment of minoritarian projects…and segregation of bodies" (Haritaworn et al., 2014, 2014, p. 6).

Assemblages of violence engage necropolitics and queer necropolitics in that they are knotted with the many ways that systems of aggression are imbricated with the politics of death, maiming, and killing. Assemblages of violence are also central to how these systems are responsible for ontological death. As participants in both research contexts experienced different yet overlapping forms of violence, who they are in the world was formed and informed by the assemblage of violence. Again, this is not to say that they lacked agency. Rather, it is a way of paying attention to the many ways that violence is "in the water" of their lived experiences. While participants in both studies discussed physical abuse (predominantly sexual assault for the women in India and being beaten by classmates or family members for the students at New Horizons), they also discussed a kind of ontological death.

Here, my argument is similar to those of scholars like Love (2019), who discusses spirit murdering of Black children. In these studies, who marginalized people can "be" in the world becomes so tied to normalized aggressions that they often express changing their ways of being to "fit" the systems of control where they live and learn. Indeed, most people are forced to adapt to social conditions in order to negotiate the world. Although presented here through violence, these conditions are not always negative; sometimes there are demands for care, empathy, and kindness. But when harmful aggressions are normalized, oppression becomes business-as-usual and becomes one's ontology, or, as Love (2019) argues, one's "spirit" becomes maimed or killed. The women at Aasha shelter, for example, often described modifying their behavior for their husband's and family's benefit in ways that became normal to them and their children over time. Not unlike the frame of living "in the closet", students at New Horizons and the women in India often closeted their ways of being for their emotional and physical safety. This is not an argument that being in a violent relationship is the same as living a closeted life. It is, however, an attention to the ontological trauma that continually shapes one's ontoepistemology in contexts of oppression.

Affects like shame are therefore both what an assemblage of violence *is* and what it *does*. These affects spill out of and then reterritorialize the assemblage within structures of power and control. Toward that end, this chapter leaves you with two more narratives that, despite the distance across which they were recorded, mirror one another in their recounting of ontological trauma, the way shame crashes on bodies, and, central to this book, how the assemblage of violence territorializes worlds.

New Horizons: Juan	**Aasha: Maaike**
What's good is that this place [New Horizons] allows you to be better than they expect you to be. And we need that, right? We all need a second chance. We need second chances. But then in that chance you learn that you're not really nothin'. You try to be better but, how can you? You'll always be what they expect you to be. Might as well just be that then, right? Me? I'll always be seen as some marica (f*ggot).	It's difficult. Sometimes I think I'm better. That in this place I've been able to get stronger, to change. I need to be able to start over. But being here also makes me feel ashamed. No one is telling me I'm worthless and yet I can't seem to feel my own value. How can you move forward when you can't even be more than what they (your family and community) expect of you? I'll always, in some way, be his.

Notes

1 It should be noted that racial hierarchies are not the only social structures in broad and, too often, pointed circulation. For example, love, compassion, and care also exist in and across systems.

2 Sati—the ritual of burning oneself alive while grieving the loss of a husband—is still practiced in some places in India. Homicides that mimic this ritual also occur, making it difficult in some cases to distinguish suicides from homicides. With the rate of domestic violence continuing to soar across the country, the numbers of suicides among women remains, from a certain perspective, anticipated yet indelible.

3 This conversation was reported to the administration at the school and, to my knowledge from student reports, the students' concerns were dealt with immediately.

Refer-ences

ActionAid. (2018). More than one in three UK women face period stigma. Retrieved from: https://www.actionaid.org.uk/latest-news/more-than-one-in-three-uk-women-face-period-stigma

Agar, M. (2004). We have met the other and we're all nonlinear: Ethnography as a nonlinear dynamic system. *Complexity*, *10*(2), 16–24.

Ahmed, S. (2010). Happy objects. In M. Gregg, & G. J. Seigworth (Eds.), *The affect theory reader* (pp. 29–51). Durham, NC: Duke University Press.

Anand, K. J. S., Rovnaghi, C. R., & Rigdon, J. *et al.* Demographic and psychosocial factors associated with hair cortisol concentrations in preschool children. *Pediatr Res* (2019). https://doi.org/10.1038/s41390-019-0691-2

Armstrong, G., & Vijayakumar, L. (2018). Suicide in India: A complex public health tragedy in need of A plan. *The Lancet Public Health, 3(10),* 459–460.

Austin, R. (2017). Gravitational Time Dilation Derived from Special Relativity and Newtonian Gravitational Potential. *European Scientific Journal*, *13*(3), 447–454.

Barad, K. (2007). *Meeting the universe halfway: Quantum physics and the entanglement of matter and meaning*. Durham, NC: Duke University Press.

BBC News. (2019). Japanese store 'rethinks' badges for staff on periods. Retrieved from: https://www.bbc.com/news/world-asia-50597405

Bennett, J. (2010). *Vibrant matter: A political ecology of things*. Durham, NC: Duke University Press.

Berger, M.. (2019). 2019 in review: A roller coaster ride for women's rights and gender equality around the world. *The Washington Post*. Retrieved from: https://www.washingtonpost.com/world/2019/12/23/review-roller-coaster-ride-womens-rights-gender-equality-around-world/

Berlant, L. (2011). *Cruel Optimism*. Bedford, O., & Hwang Durham, NC: Duke University Press.

Bohn, D. K., & Holz, K. A. (1996). Sequelae of abuse: Health effects of childhood sexual abuse, domestic battering, and rape. *Journal of Nurse-Midwifery*, *41*(6), 442–456.

Bourassa, C., McKay-McNabb, K., & Hampton, M. (2004). Racism, sexism and colonialism: The impact on the health of aboriginal women in Canada. *Canadian Woman Studies*, *24*(1).

Bourdieu, P., & Passeron, J. C. (1977). *Reproduction in education, society and culture*. London, UK: Sage Publications.

Brewin, C. R. (2015). Re-experiencing traumatic events in PTSD: New avenues in research on intrusive memories and flashbacks. *European Journal of Psychotraumatology*, *6*(1), 27180.

Britzman, D. P. (1995). Is there a queer pedagogy? Or, stop reading straight. *Educational theory*, *45*(2), 151–165.

Daphna-Tekoah, S. (2019). My body protests: Childhood sexual abuse and the body. *Journal of loss and trauma*, *24*(5–6), 533–549.

Das, M. (2018). 'Drunk bloques and suicidal women more dangerous than the ocean, *The Times of India*, n.p., Retrieved from: https://timesofindia.indiatimes.com/india/drunk-blokes-suicidal-women-more-dangerous-than-ocean/articleshow/66781494.cms

DeLanda, M. (2016). *Assemblage theory*. Edinbrugh, UK: Edinburgh University Press.

Deleuze, G., & Guattari, F. (1980). *A thousand plateaus: Capitalism and schizophrenia*. London: Continuum.

Deo, D. S., & Ghattargi, C. H. (2005). Perceptions and practices regarding menstruation: A comparative study in urban and rural adolescent girls. *Indian J Community Med*, *30*(1), 33–4.

Derrida, J. (2008). *The animal that therefore I am*. New York, NY: Fordham University Press.

Douglass, F. (1855). *My bondage and my freedom*. New York, NY: Modern Library.

Du Toit, B. M. (1988). Menstruation: Attitudes and experience of Indian South Africans. *Ethnology*, *27*(4), 391–406.

Fahs, B. (2016). *Out for blood: Essays on menstruation and resistance*. New York, NY: SUNY Press.

Feagin, J. R., & Van Ausdale, D. (2001). The first R: How children learn race and racism. Landham, MD: Rowman & Littlefield Publishers.

Foucauldian (1978). *The history of sexuality*. New York: Pantheon Books.

Foucault, M. (1976). *The history of sexuality, volume 1*. New York, NY: Pantheon Books.

Geertz, C. (1973). *The interpretations of cultures*. New York: Basic Books.

Gershon, W. S. (2017). *Sound curriculum: Sonic studies in educational theory, method & practice*. New York: Routledge.

Gershon, W. S. (2018). Reverberations and reverb: Sound possibilities for narrative, creativity, and critique. *Qualitative Inquiry*, 1–11.

Grace, J. E., & Nelson, S. L. (2019). "Tryin'to survive": Black male Students' understandings of the role of Race and racism in the school-to-prison pipeline. *Leadership and Policy in Schools*, *18*(4), 664–680.

Gregg, M., & Seigworth, G. J. (Eds.). (2010). *The affect theory reader*. Durham, NC: Duke University Press.

Halberstam, J. (2009). Queer studies. In P. Essed, D. T. Goldberg, & A. Kobayashi (Eds.), *A companion to gender studies* (pp. 62–72). Malden, MA: Blackwell Publishing Ltd.

Halperin, D. M., & Traub, V. (Eds.) (2009). *Gay shame*. Chicago, IL: University of Chicago Press.

Haritaworn, J., Kuntsman, A., & Posocco, S. (Eds.) (2014). *Queer necropolitics*. New York, NY: Routledge.

Hartman, S. V. (1997). *Scenes of subjection: Terror, slavery, and self-making in nineteenth-century America*. New York, NY: Oxford University Press.

Helfenbein, R. J. (2010). Thinking through scale: Critical geographies and curriculum spaces. In E. Malewski (Ed.), *Curriculum studies handbook: The next moment* (pp. 304–317). New York, NY: Routledge.

India Today (2019). Sabarimala temple opens today, no protection for women activists. Retrieved from: https://www.indiatoday.in/india/story/sabarimala-temple-opens-today-no-protection-to-women-1619503-2019-11-16

Jain, C. R. (1915). *The key of knowledge*. New Delhi, India: Today & Tomorrow's Printers & Publishers.

Johnson, E. P. (2016). *No tea, no shade: New writings in black queer studies*. Durham, NC: Duke University Press.

Kawada, M. (2019). *Origin of ethnography in Japan*. New York, NY: Routledge.

Kohn, E. (2013). *How forests think: Toward an anthropology beyond the human*. Oakland, CA: University of California Press.

Kohut, H. (1972). Thoughts on narcissism and narcissistic rage. *The Psychoanalytic Study of the Child*, 27, 360–399.

Lindisfarne, N. (1998). Gender, shame, and culture: An anthropological perspective. In P. Gilbert & B. Andrews (Eds.), *Series in affective science. Shame: Interpersonal behavior, psychopathology, and culture* (p. 246–260). Oxford University Press.

Love, B. L. (2019). *We want to do more than survive: Abolitionist teaching and the pursuit of educational freedom*. Boston, MA: Beacon Press.

Massey, D. (2005). *For space*. Thousand Oaks, CA: Sage Publishing.

Mbembe, A. (2019). *Necropolitics*. Durham, NC: Duke University Press.

Mbiti, J. S. (1969). *African religions and philosophy*. Portsmouth, NH: Heinemann.

McKittrick, K. (2006). *Demonic grounds: Black women and cartographies of struggle*. Minneapolis, MN: University of Minnesota Press.

McKittrick, K. (2015). *Sylvia wynter: On being human as praxis*. Durham, NC: Duke University Press.

Morris, M. W. (2016). *Pushout: The criminalization of black girls in schools*. New York, NY: The New Press.

Moten, F. (2003). *In the break: The aesthetics of the black radical tradition*. Minneapolis, MN: University of Minnesota Press.

Munt, S. R. (2017). *Queer attachments: The cultural politics of shame*. New York, NY: Routledge.

Nathanson, D. L. (Ed.) (1987). *The many faces of shame*. New York, NY: The Guilford Press.

Ortner, S. B. (2006). *Anthropology and social theory: Culture, power, and the acting subject*. Durham, NC: Duke University Press.

Ostle, C., Thompson, R. C., Broughton, D., Gregory, L., Wootton, M., & Johns, D. G. (2019). The rise in ocean plastics evidenced from a 60-year time series. *Nature communications*, *10*(1), 1–6.

Phipps, A., Ringrose, J., Renold, E., & Jackson, C. (2018). Rape culture, lad culture and everyday sexism: Researching, conceptualizing and politicizing new mediations of gender and sexual violence. *Journal of Gender Studies*, *27*(1), 1–8.

Parise, A. G., Gagliano, M., & Souza, G. M. (2020). Extended cognition in plants: Is it possible? *Plant Signaling & Behavior*, 15(2), 1–5.

Pasco, R. J. (2003). *Capital and opportunity: A critical ethnography of students at-risk*. Lanham, MD: University Press of America.

Puar, J. (2007). *Terrorist assemblages: Homonationalism in queer times*. Durham, NC: Duke University Press.

Puar, J. (2012). 'I would rather be a cyborg than a goddess': Becoming intersectional in assemblage theory. *philoSOPHIA*, *2(1)*, 49–66.

Rivera Antara, D.. (2019). UK pledges £2m to help end stigma and shame of period poverty. Retrieved from: https://www.theguardian.com/global-development/2019/mar/05/uk-pledges-2m-to-help-end-stigma-and-shame-of-period-poverty

Robson, D. (2020). Spider smarts. *New Scientist*, *245*(3268), 42–45.

Rosaldo, R. (1993). *Culture & truth: The remaking of social analysis*. Boston, MA: Beacon Press.

Rosiek, J. L. (2018). Agential realism and educational ethnography. In D. Beach, C. Bagley, & S. Marques da Silva (Eds.), *The Wiley handbook of ethnography of education* (pp. 403–423). Hoboken, NJ: John Wiley & Sons, Inc.

Sabzalian, L. (2019). *Indigenous children's survivance in public schools*. New York, NY: Routledge.

Sedgwick, E.K. (1993). The epistemology of the closet. In H. Abelove, M.A. Barale & D.M. Halperin (Eds.), *The Lesbian and Gay Studies Reader* (pp. 45–62). New York: Routledge.

Sharpe, C. (2016). *In the wake: On blackness and being*. Durham, NC: Duke University Press.

Skow, B. (2018). The tenseless theory of time and the moodless theory of modality. *Pacific Philosophical Quarterly*, *100*(2), 506–524.

Stockton, K. B. (2006). *Beautiful Bottom, beautiful shame: Where "Black" meets "queer*. Durham, NC: Duke University Press.

Sutton, H. (2016). Transgender college students more at risk for suicide when denied bathroom, housing rights. *Campus Security Report*, *13*(2), 9–9.

Synnott, A. (1987). Shame and glory: A sociology of hair. *The British journal of sociology*, *38*(3), 381–413.

Thiagarajan, K.. (2019). Millions of women in India join hands to form a 385-mile wall of protest. *NPR*. Retrieved from: https://www.npr.org/sections/goatsandsoda/2019/01/04/681988452/millions-of-women-in-india-join-hands-to-form-a-385-mile-wall-of-protest

Tsing, A. L. (2005). *Friction: An ethnography on global connection*. Princeton, NJ: Princeton University Press.

Valenzuela, A. (1999). "Checkin' up on my guy": Chicanas, social capital, and the culture of romance. *Frontiers: A Journal of Women Studies*, *20(1)*, 60–79.

Walkington, C., Chelule, G., Woods, D., & Nathan, M. J. (2019). Collaborative gesture as a case of extended mathematical cognition. *The Journal of Mathematical Behavior*, *55(1)*, 100683.

Weheliye, A. G. (2014). *Habeas viscus: Racializing assemblages, biopolitics, and black feminist theories of the human*. Durham, NC: Duke University Press.

Wiesel, E. (2002). *Elie wiesel: Conversations*. Oxford, MS: University of Mississippi Press.

Woodson, C. G. (1933). *The miseducation of the Negro*. Washington, DC: The Associated Publishers.

Wozolek, B. (2019). M/othering redoux: A Reading of curriculum theory to get through sleepless nights. *The Journal of Gender Studies, 28(8),* 873–882.

Yehuda, R., & Bierer, L. M. (2009). The relevance of epigenetics to PTSD: Implications for the dsm-v. *Journal of traumatic stress*, *22*(5), 427–434.

Yosso, T. J. (2005). Whose culture has capital? A critical race theory discussion of community cultural wealth. *Race ethnicity and education*, *8*(1), 69–91.

3

PATHOLOGY AS METHOD

Tracing Trajectories

Sample 1: Happiness and Chosen Kin

KATIE: *She came!*

WOZOLEK: *Who?*

KATIE: *My girlfriend! She came! She cares about me so much that she came to my presentation at the assembly on Langston Hughes. We were able to invite family so I asked her to come. Like she cut school, put on her old uniform from eighth grade so no one would notice, and came to the school. My mom can't come. Well, she won't come. She doesn't do anything for me since I came out. But Bryiann, she's my family anyway. I have to admit, I was hoping she would be there and then she was there! You know what I like about her? I think I like that she makes me feel like I'm not a failure. Everyone here sees me as one. But she doesn't. She just sees me. And that makes me happy.*

WOZOLEK: *What else makes you feel happy?*

KATIE: *I don't know. I remember once a teacher told me this old quote from an old movie about happiness being in your backyard or something. I don't know. What if I can't find happiness? What if happiness has to find me? Well, it did. Bryiann did.*

Everyday affects in our lives are always already complex, from the moment they are experienced in one body until the moment they stick on, between, and through other beings. Like a kaleidoscope that is changed as much by internal movements as by the rotation of the loose cell, people are as impacted by their reaction to affects as they are by the sociocultural and political structures in which affects are constructed. Kathleen Stewart (2007b) described this complexity by stating that affects can "be experienced as an empty pause or a dragging undertow, as a sensibility that snaps into place or a profound disorientation. They can be funny, perturbing, or traumatic" (Stewart, 2007b, p. 2). The pervasive nature of affects across layers of experience has sparked dialogues on affective relations from the social sciences (e.g., Gregg & Seigworth, 2010, 2010; Skattelbol & Hayes, 2016; Tompkins, 1962) to physical science (e.g., Fosha et al., 2009, 2009; Wild, 2005). Though there are multiple inroads to these complicated conversations, this chapter focuses on a method for tracing one ordinary affect (Stewart, 2007b): happiness.

Using "pathology" as methodology, this chapter traces happiness as it simultaneously erupts and quietly flows through the lives of research participants, across their contexts and within assemblages of violence. Historically, "pathology" has been a term exclusive to the medical community, often describing the diagnosis of a disease or illustrating predicted progressions, and other times attending to paths that deviate from an expected course (Robbins, 2010). A pathology is traditionally read as a process that details a linear progression (see, for example, pathologic conditions that impact the body or the mind). Pathology, however, can be also used to express *any* causes, effects, progressions, or pathways (e.g., Bartlett, 2005; Werner & Benjamin, 2002).

To show how pathology can be used to approach empirical data, this chapter begins with the vignette above to walk through the steps of a qualitative pathology, or to borrow from Feltham (2007), an anthropathology. Following this, I discuss some of the histories of pathology as it exists in the medical community, as well as some of the complications of claiming what is often perceived as a quantitative medical tool to discuss qualitative research. Finally, using several other data points and applying pathology as method, I show how pathology can be applied practically across contexts.

Anthropathology: The Mechanics of Tracing

While the opening to this chapter might generally be understood as a vignette, I have recast it as a "sample". The use of the word "sample" here is intentional and, admittedly, potentially problematic, because it can be understood as a return to the medical language from which qualitative research has worked to extricate itself for generations, for many ethical reasons (Denzin & Lincoln, 2008). While I could use the term "case", I have decided to use "sample" for the following reasons. First, it is always debatable what makes a "case" a case (Yin, 1981). Ultimately, a "case" is an artificial set of understandings around some phenomenon. A "sample", on the other hand, is a portion of something taken for the explicit reason of examining a quality or characteristic of that thing. Here, I have taken an iteration of happiness and, by examining it with pathology in mind, I trace its pathways. In addition, using the term "sample" is helpful because it is broad enough to encompass several possibilities in terms of what is collected, but narrow enough to denote that one "thing" is being studied.

Second, although the framing of "sample" remains helpful, my use of it here rejects associated positivist concepts such as validity and generalizability, and therefore allows for qualitative analysis without remaining beholden to quantitative research. Finally, when one sample renders features similar to another, as happens in much qualitative research, this can allow for comparison without generalization, helping us better understand contexts and, perhaps, allowing us to think about how certain injustices might be interrupted. My use of the word is, in essence, a reappropriation of the idea of the sample.

I chose Katie's sample because, on initial examination, it might appear as though she has found happiness despite the complications of her life. However, in this instance, as following a method of pathology will document, Katie's happiness is still enmeshed with violence. This is why additional context is significant. Whether it is an anthropologist collecting field notes or a pathologist attending to patient history, context helps us understand how a sample moves or changes in connection to other people, spaces, and places. It is also because of context that we must attend to biases, both of the speaker and of the observer.

Prior to this sample, Katie was failing several classes. The time during which I met with students was considered an "enrichment" period, meaning that they could either participate in extracurriculars like the Blue Dragons or receive tutoring. Katie was considered an "underperforming" student, so she often had to skip time with the Blue Dragons to receive one-on-one tutoring. She had also been kicked out of her house for a week when she first came out, but was then asked to return when, as Katie put it, her mom could "handle herself" more. She had not been looking forward to presenting at the Langston Hughes assembly because, as she described, she didn't "want everyone to see [her] and know that [her] work [was] shitty". She also had doubts about Bryiann, who had moved on to another school. This meant their time together was limited to nonexistent. However, her mother, who was aware of her feelings for Bryiann, had been more "relaxed" since she knew that Bryiann was no longer at New Horizons. This meant that she was generally "more happy and kind" to Katie.

Katie's sample gives the reader some points of comparison, most notably between Bryiann and her mother, and a second point between the quote from *The Wizard of Oz* and Katie's own thoughts about happiness. While the practice of pathology tends to negatively pathologize one thing as abnormal in comparison to another, I use it here not to claim abnormality but rather to ethically attend to histories, contexts, and comparisons as a means to disrupt normative cis-heteropatriarchal values. At the root of Katie's narrative is a tension between these points of comparison that are centered on happiness. Katie, who has experienced violence in school and at home, is content in her happiness, yet the tension is still palpable—while Bryiann showed up this time, Katie is still nervous, wondering if Bryiann will be able to attend future events. As a result, she tacks back and forth between her relationship with Bryiann and how others might interpret happiness. As I explain below, Katie's sense of happiness is akin to Ahmed's (2010) and Berlant's (2011) scholarship, which argues for an understanding that happiness exists in connection to, and in many cases can reinforce, violence.

The point of the sample, then, is to trace how happiness has moved in connection to violence. At the same time, I use this sample to document how pathologies can function methodologically. This method is not necessarily meant to be predictive, especially as that might provoke self-fulfilling prophecies. It is, however, an opportunity to think about how affects like happiness move in relationship to broader aggressions, even if these are not foregrounded in the sample.

Tracing Beginnings

Before I began to articulate Katie's narrative as a sample, I was vexed by the affects that existed everywhere and nowhere at once within the data. Specifically, the movement of affects remained incorporeal yet perceptible throughout my work at both New Horizons and Aasha. I began to note how affects seemed to be drawn to each other at some times and at other times repelled. Affects, it seemed, created their own forces. And as these energies pushed and pulled them across and sometimes out of assemblages, I sometimes lost track of affects as they shuttled (Gregg & Seigworth, 2010, 2010) toward and away from participants and myself. I first began to theorize these movements through the metaphor of magnetism, but I was unconvinced that the binary of magnets—positive and negative poles—was helpful in describing the narratives of queer youth of color and women who were survivors of intimate partner violence. As I described in the preface, it was not long before I arrived at quantum entanglements to explain one of the ways that affects travel within an assemblage. But this did not explain how one might trace the complex movements of affects, what Gregg and Seigworth (2010 (2010) describe as "intensities".

About the same time that I was working through these questions, I found myself sitting at a hospital in Seattle holding the hand of my best friend, Beth, someone I had loved for 22 years. We received the devastating news that her cancer was widespread and had a genetic mutation that made it difficult, if not impossible, to treat. I remember listening to the doctor and scribbling down new words as quickly as I could so that I could look them up later to better understand the diagnosis—perhaps with some false hope that I could research my way into a treatment plan that the doctors hadn't yet considered. I remember waiting each evening until Beth's mother left and she had fallen asleep to return to these notes, pulling up all the medical journals available from my institution. I looked up new terms and applied them to various texts long into the night, as nurses periodically interrupted Beth's sleep and my futile inquiries into oncology. At some point in my emotional and physical exhaustion, in the middle of reading a medical journal, I arrived at the answer to my work at New Horizons and Aasha—pathology.

The more I considered the connection, the clearer it became that pathology was one way to trace the ever-emergent affects I had been struggling with. Not only because pathology would allow me to look at things that might be malignant, like violence, or benign, like happiness or joy, but also because the act of tracing something through pathology implied a degree of care and attention: The care of deciding which sample will be useful. The attention is to how, from whom, where, and when a sample is taken. The intention to interrupt something that is harmful. In this moment, I was able to see more clearly how happiness was central to the lives of participants yet entangled within the assemblage, and I could envision a way to trace its movements. Perhaps most significantly, I saw how my training as a qualitative researcher was nested and layered with the ways that pathology might be used to trace affects, rather than existing separately from it. Rather than lingering on how I landed on pathology, in the next section I discuss some of the theorization behind pathology as both concept and method. I also frame the idea of anthropathology, as it is used here, before returning to anthropathology as a method to trace happiness and other samples.

Tracing the Path

The recent COVID-19 pandemic offers an example of how researchers work tirelessly to trace both the pathology of a virus as it ravages bodies, and the pathways that the virus takes as it spreads through communities across the globe. Although accompanying understandings of pathology can be negative to the point of feeling fear, the "thing itself" is viewed objectively. For example, cancer is a negative diagnosis in terms of what doctors, patients, and families alike hope for in terms of test results. Although the prognosis is not always abysmal, sometimes indicating a low chance of mortality, the diagnosis generally elicits understandably negative feelings like fear. Yet the cancer can also be regarded as something that simply exists, devoid of any connotation. From this perspective, pathology is a tool for tracing the trajectories of movement; it unveils the object of study, the malignancy or benign nature of which is a secondary characteristic.

Determining something's path can also be understood as creating a cartography of affects. Not unlike Deleuze and Guattari's (1987) tracing of a rhizome, Massey's (2005) discussion of space, or Helfenbein's (2010) conversation on sociocultural bleeding between spaces, a cartography of affects is complex and fluid. Pathology offers one way to understand the places and spaces created by affects as they travel through an assemblage and, perhaps more importantly, how we might think about future becomings as the affects continue to travel. The purpose of tracing these movements is not to further marginalize people through the construction of self-fulfilling prophecies, but rather to think about how everyday oppressions can be interrupted.

Pathology offers a way to attend to the materiality of qualitative research without losing the emphasis on fluidity that is central to scholarship using rhizomes as metaphor or with rhizomatic findings. Dialogues focused on rhizomatic relationships do the painstaking work of showing the wonderful, complex nature of events, ideas, and ways of being, knowing, and doing (e.g., Deleuze & Guattari, 1987; Snaza et al., 2016, 2016). However, this scholarship tends to focus on what a rhizome or assemblage is, or occasionally on what it does, but with little discussion of how it moves beyond noting its complexity and seemingly random nature. What is generally missing is how one might practically trace the movements of everyday affects. Pathologies are helpful because they trace such fluid movements without reducing their complexity or immediately demanding value judgments.

Although not central to my use of pathology in this chapter, there are two aspects of the concept I briefly address here; without such attention, these could potentially be harmful and marginalizing. First, the use of pathology has historically focused on harmful agents. In contrast, I use pathology as method—or as the path of the method, process, or study—to trace patterns and how they function without positing that the events, bodies, or beings involved have negative connotations. In this book, where queer youth of color and women who are survivors of domestic violence are the primary participants, it is important to be clear that by tracing experiences through pathology I do not position the participants, their experiences, or their ways of being and knowing within the traditional use of the term, which could carry negative framing or eugenic tendencies.

Second, pathology has been used to study everything from psychological progressions to physiological changes with some degree of objectivity. But objectivity is impossible and, as discussed in detail below, is not the purpose of this exercise. As a queer woman of color who has faced emotional and physical violence, it is impossible for me to trace the trajectories of happiness in relation to these topics without entangling my own vulnerabilities and bias (Behar, 1996). This is true of every researcher, whether she be an anthropologist, a pathologist, or, as I discuss below, an anthropathologist (Feltham, 2007). Historically, researchers tried to act as though bias was not implicitly and explicitly ingrained into their ways of being, knowing, and doing. More specifically, atrocities have occurred when researchers slipped between following a pathology objectively and transferring that objectivity to interactions with a person. One only needs to think about the Mengele Trials, the Tuskegee study, the Sims surgical experiments, the monster study, or Japan's 731 Unit to see how medical inquiry has been used across the globe to value pathology over people.

Physical sciences are not the only place where oppressions have occurred due to overconfidence in false objectivity. Long-standing critiques of objectivity can be heard across the social sciences, including in critical race scholarship (e.g., Ladson-Billings, 1998), feminist scholarship (e.g., Lather, 1991; Miller, 2005), queer theory (e.g., Halberstam, 2011; McCready, 2004), and critical qualitative research (e.g., Denzin, 2017; Noblit, 2004). These complicated conversations (Pinar, 2011) are both about resistance to objectivity and rejection of positivist norms within the scientific method and about nesting subjectivities within the critical distance of qualitative research. Such scholarship is extremely important, but rather than rejection I propose a continued attention to ways that objectivity has been, and continues to be, used as a tool of oppression.

In agreement with these critiques, I do not argue for pathology as an objective method. As strong qualitative research has well observed, a pathology may be neutral but those analyzing it are never without subjectivities (Rosaldo, 1993; Tsing, 2005). Instead of seeking objectivity, I follow Deleuze and Guattari's (1987) theorization of rhizomes as a way to describe the multiplicities, complications, and entangled webs of relations that form and inform ways of being, knowing, and doing (e.g., Barad, 1999; Massumi, 2002; Snaza et al., 2016, 2016; Weheliye, 2014). Deleuze and Guattari argue that the rhizome exists in multiplicity because of its twinned nature. On one hand is a side "signifying totality, or determination attributable to a subject...[and, on the other] a side which is continually dismantling the organism, causing asignifying particles or pure intensities to pass or circulate, and attributing to itself subject that it leaves with nothing more than a name as the trace of an intensity" (p. 4). Here, Deleuze and Guattari explicate how rhizomes are ethereal but never ahistorical, and, when traced with the affective tensions of people, are also never apolitical or acultural.

Similar to their tracing of rhizomes, I use pathology as a method to trace affects, such as the happiness elicited in the samples from these studies.

While Deleuze and Guattari's theorization of entanglements is significant, this work often misses materiality when applied to new contexts or questions. Pathology is one way to be clear about the deeply concrete, material implications and nature of entanglements (Helfenbein & Huddleston, 2013). Additionally, following dialogues on the necessity of fluidity in qualitative research (e.g., Denzin, 2017; Gershon, 2017a; Rosaldo, 1993) and strong social theory (e.g., Ahmed, 2010; Lather, 1991; Snaza et al., 2016, 2016), this method uses scientific terms from the turn of the last century. The method itself follows critical dialogues in education, qualitative research, and social theory (e.g., Anyon, 2008; Evans-Winters & Esposito, 2010; Ibrahim, 2015; Noblit, 2004; Ortner, 2006; Pillow, 2015) that reject what might be read as a positivist framing.

As but one example of these complications is the term "anthropathology" (Feltham, 2007), a phrase I use relatively interchangeably with pathology throughout this chapter. Feltham (2007) fleshes out this neologism in his discussion of violence. Prior to Feltham's scholarship, the term "anthropathology" had a rather oppressive history within the United States. Specifically, it was often used as a diagnosis to declare people themselves, particularly African Americans, as a form of "degeneration and disease... through a discursive fusion of anthropology, medicine, and genetics" (Tapper, 1997, p. 263). The term was often used for sickle-cell patients, and this diagnosis was used within eugenicist communities as an exclusionary tool through medicine. Feltham acknowledges this history but uses the word to discuss how humans tend toward paths of violence that, when entangled with other oppressions, make our sociopolitical and cultural norms untenable. Feltham does, however, stay with pathology as a way to document the "breadth of 'humanity' and the weight of 'sickness' to... examine the more miserable, pessimistic view... of our human character" (p. 2–3). Feltham's use of anthropathology therefore considers both past and present pathways of violence in and across communities; he explicates how violence moved and is currently moving.

Building on Feltham's work, I use anthropathology not only to trace past and present affective movements but also to think about the future trajectories of bodies and their affects. This is not an attempt to predict the future—something that all too often ends in self-fulfilling prophecies for marginalized people—but rather an attempt to think about how oppressive norms might be interrupted. As with pathology in modern medicine, movements are traced not only to map the current cartography but also to think about how things might move in the future.

Rather than being inherently positivist, then, pathology as method is inherently fluid as I apply it to the intra-actions of and between bodies. The "sample", then, is an inroad to thinking about the knotted and complex nature of intra-actions within an assemblage. By using anthropathology, I address nonlinear or sequential rhizomatic relations and observe the fluidity of connections and contexts, all of which have a multiplicity of trajectories and are omnidirectional in their movements.

Finally, pathology is also like strong qualitative research in that analysis of contemporary events is always already historically connected. Not unlike field work, pathology starts with observation of a context. However, once a researcher leaves their place of study, the data is historicized, as the broader community continues to move after the researcher has left. Similarly, within medical pathology, a sample is taken and studied closely to understand possible expressions, changes, and paths of progression. Yet the subject from whom the sample was taken continues to develop. In short, all qualitative research involves efforts at pathology. However, pathology departs from some traditional qualitative research in that human interactions do not necessarily have to be the focus and concrete understandings are fluid.

This chapter now turns to a theoretical discussion on happiness, using pathology as described above. All three samples included in this chapter attend closely to happiness as it moves within an assemblage of violence. The samples were also similar in that they all contained insights into chosen family as a space that engendered happiness. An inherent tension exists at the nexus of happiness, chosen family, and violence; I explore this tension below, first through a theorization of happiness and then again by applying pathology as method to the samples.

Happy Relations: The Optimism of Chosen Families

Writing about happiness is not new. For example, nihilistic critiques of happiness in relation to hope are common, particularly within post-Kantian philosophy (Camus, 1938; Nietzche, 1878). Further scholarship captures different aspects and engagements of happiness (Ahmed, 2010; Berlant, 2011; Dewey, 1916; Halberstam, 2011). Finally, related to the stories of students like Katie at New Horizons, some scholars have theorized happiness as it is related to hope through educational dialogues (e.g., Freire, 1992; Gershon, 2015; Giroux, 1997; Helfenbein, 2004; Snowber, 2013; Taliaferro Baszile & Edwards, 2016; Wrigley et al., 2012). Building on these works, my point here is to show how tracing the pathology of happiness can function in and against systems of oppression.

Happiness is understood in these scholarly examples as rhizomatic, in that is has no master plan and exists between and within intensities. Further, it has no set archetypal structure yet is inherently interconnected and—depending on particular actions, contexts, and engagements—nonhierarchical. Within this fluidity, happiness can move people and events—taking them down paths or perhaps in circles, or to unexpected destinations. It also exists in multiplicity with the property of asignifying rupture (Deleuze & Guattari, 1987), meaning that happiness is constantly deterritorializing only to re-form within new identities and possibilities. Or, as Moten (2003) describes, happiness functions not only as a significant event that *is* but also as a marker for changes to come. Although happiness can have multiple entry points, and those experiencing happiness can never truly relive an experience from the same entry, this chapter traces lines of flight through relationships, both familial and chosen.

Happiness also has movement in that it changes across layers or scales. For example, the act of finding or creating happiness can push against systems of oppression that are not only emergent but also sometimes designed to promote sorrow. In this sense, pathology can be a tool to leverage against oppressive ideas and ideals; this is the agency of affect that I discussed in the previous chapter. On the other hand, the desire of an oppressor to engender and maintain lines of power is the opposite side of that coin. Happiness moves through its presences and absences and, because of this movement, has historically been difficult to trace. Pathology, the mapping of pathways and trajectories, offers a possible solution to this long-standing complication.

The critical examination of happiness is important because, just like hope, it is always already entangled with violence. While Katie's sample may be read as a moment of care between two African American middle school girls, it is also nested within questions of budding sexualities, tensions within queer Black communities, self-fulfilling prophecies centered on academic achievement, and chosen familial ties, to name a few. In the middle of such complications, it is important to think deeply about how affects like happiness are entangled with/in the daily lives of students like Katie and Bryiann. Similar to Ahmed's (2010) discussion of the stickiness of happiness and Berlant's (2011) scholarship on cruel optimism as an inversion of what is often conceptualized as queer pain and ostracization, reflecting on these relationships is a way of paying attention to the local conditions under which one experiences happiness. At the same time, this approach also focuses on the less-local entanglements between sociopolitical ideas and ideals about happiness.

Unlike joy, often discussed as internally created or a "liquid that fills the body" (Tseng et al., 2005, p. 361), happiness is typically conceptualized as existing in relation to other bodies (e.g., Ahmed, 2010; Berlant, 2011; Cottrell, 2016; Tseng et al., 2005). As Ahmed (2010) argues, happiness is often conditional on interrelated objects and people, which are anticipated to cause happiness. She suggests that happiness often takes the form of a promise, of deferred possibilities rooted in the idea and ideal of future comfort. Happiness, Ahmed explains, is often thought of as a forward orientation that becomes an injunction to put negative experiences behind the person seeking to be happy or content. Should the person seeking happiness not find what he is searching for, the onus for the absence of happiness is ultimately placed on that person, not on all-too-often marginalizing cultural norms and values. This means that an oppressed individual is then blamed for not finding or creating the conditions under which she can feel content. In Katie's sample, she refers to a quote from the movie *The Wizard of Oz* that states, "If I ever go looking for my heart's desire again, I won't look any further than my own back yard. Because if it isn't really there, I never really lost it to begin with" (Baum et al., 1939). Her concern underscores Ahmed's points about happiness—a fear that happiness is both unattainable and under her control.

Happiness is also considered a moral imperative. The "duty" to be happy is a social responsibility to live one's life in a way that makes others—parents, partners, teachers, and community members, among others—feel happy in relation to you (Ahmed, 2010). In both studies, participants were called toward an ideal of happiness that was shaped by violent norms and values. For example, many women in India expressed that their happiness was centered on their husband's ability to be content. Similarly, many students at New Horizons expressed that antiqueer values left them wondering if they would be happier in the closet, especially as it would make their parents, classmates, and teachers feel more "comfortable". Used as a tool for violence, happiness functions as an agentic contingency whereby both the women's and students' happiness was a point of relation, an intra-action, between themselves and broader oppressive entanglements. Their agency within happiness was limited by and through violent values. However, happiness can also function as an act of agency that presses back against assemblages of violence.

Katie's feelings for Bryiann are in many ways akin to Berlant's (2011) discussion of cruel optimism, because of antiqueer norms and values, Katie's desire for Bryiann is an obstacle to her flourishing in a heteronormative society. Berlant argues that, because people remain attached to fantasies of the "good life", there is a relational dynamic in which people seek out "conditions of possibility" or "clusters of promises that are rooted in desires, even when such wants inhibit the conditions for fulfilling such promises" (p. 24). Berlant uses obesity as but one example of cruel optimism. Obesity, Berlant writes, is a form of slow death through the destruction of healthy organs, particularly for those whose food options are limited to high-calorie, nutritionally devoid foods. In Katie's case, the normalization of queerphobias set up an "affective impasse, a set of felt relationships that cannot be moved through" (Schaefer, 2013, para. 4). *Cruel Optimism* attends to such impasses, exploring where and how desires emerge, whether in response to trauma or to normalized understandings of other affects, like happiness.

The distinct yet interrelated theories of Berlant and Ahmed offer a space to think about the relationship between bodies and discourses, central not only to affect theory but also to the entangled intra-actions that are always present within assemblages. Affects are not possible without bodies to bounce between or to send, receive, and create them. While the bodies that construct affect are not necessarily human—otherwise there would not be such a plethora of happy animal videos on the internet—bodies do need to be in relation to something for affects to occur. Whether it be a dog jumping into a lake with wild abandon, a baby laughing at her family pet, or a fox enjoying the sunlight, affects occur through relations. In regard to affect and the knotted nature of assemblages, there tends to be a great deal of focus on the knot itself and what that knot does in and across contexts (e.g., Dernikos et al., 2020; Gregg & Seigworth, 2010, 2010; Massumi, 2002; Puar, 2013; Weheliye, 2014). I have raised this point several times throughout this book, from the quantum entanglements that move others as much as they move, to the assemblages that territorialize sociocultural and political norms and values, to what we learn through and because of an assemblage. Here, however, I ask how we might trace affects to think about how they *will* function, rather than how they *have* functioned.

While happiness can be traced through multiple sites, the samples I chose from the two studies focus specifically on chosen family. For the youth at New Horizons, such kinship often signified the cluster of "promises they wanted someone or something to make possible for them" (Berlant, 2011, p. 23). The significance of such relationships across cultures and, in resonance with this study, for queer people of color, has been noted in a variety of places, from scholarly works (McCready, 2004; Stewart, 2007a) to conversations in pop culture. One illustration is Billy Porter's recent interview on *The Late Show With Stephen Colbert*, in which he discussed his role on the show *Pose*. Porter passionately noted that 1980s ballroom culture was important, especially to Black queer men, as "a culture of chosen family". Porter continued, "Sometimes our biological families are not equipped to love us unconditionally in the ways that are necessary for us to thrive when we are LGBTQ people". Similarly, I use the data below to articulate how such relationships are central to the participants. Although this data was a part of a larger study, pathology begins by lifting one central moment of happiness—the sample—from each of the studies and analyzing their relations to the broader context.

As with any pathology, the sample provides enough substance to discuss the underlying conditions of the context, but it is not a form of the thick description (Geertz, 1973) that is central to much qualitative research. While this might seem antithetical to qualitative research histories that value clear, detailed descriptions as a form of transparency, the reality is that all qualitative research leaves ideas on the cutting room floor (Agar, 1996), providing only a sample of the context for the readers. My purpose here is, on one hand, to be clear about the cuts that I make in my research and, on the other, to use a sample to trace movements that were not only central to these narratives but also representative of the broader studies.

"Fuck You" Happiness: José's Sample

The following sample focuses on José, a first-generation seventh-grade Mexican American student who identified as questioning his sexual orientation. José was quiet, often sitting on his own during group meetings. He expressed that his grandfather was incredibly upset when he heard that José was thinking about kissing another boy at school, a secret that he had confided with his cousin. As a result of his questioning identity, he explained, he had difficulty maintaining relationships with his blood relatives. He described his mother as being "close to tears" since she had found out about his desires, and his grandfather kept dropping non sequiturs into their conversations about how to "be a man". These lessons were almost always accompanied by some form of emotional violence, and on a few occasions, physical violence. As a coping mechanism, José fostered relationships with classmates who were either queer or allies, and he formed a platonic relationship with a gay Latino man who lived on his block.

José explained multiple qualities that attracted him to chosen family members. But, as in the sample below, he also described that a sense of happiness was drawing him toward fantasizing about a relationship with another student, Jarod. Throughout the study, their relationship moved between romantic love and friendship, settling on a combination of deep care, love, and respect by the end of the two years. In communities and a school where homo- and transphobias are normalized, and where emotional and physical violence are rooted in racism and antiqueerness, José's ability to "feel happy and say 'fuck you' to everyone else" was exceptionally important.

Sample

JOSÉ: Happiness? Happiness is being with your "familia". No...not your "familia" like your mom and your dad and your grandpa. I'm talking about *your* family. The people who *really* take care of you. Like Jarod. Jarod and me....Well, we been together, runnin' around the barrio for, well, forever. It's hard for him because there are so many Mexicans in our part of town and he's like one of the only Black kids. Not that there aren't plenty around the city or even like three blocks over. But I'm talking about where I'm from. Some people, like my sister, like to tease me about being friends with him. Like she asks me why I can't play with the other boys "like us". I'm older now so I just tell her that's racist and she goes away. Anyway, he keeps me safe. He even got me a knife. You know, just in case, 'cause people be buggin' me when I walk to school. [pause.] Happiness… You asked me what makes me happy… So, I know I told you some of this before but Jarod. Well, in sixth grade I kept havin' this thought, when we were playin' or just goin' around the streets… I just wanted to kiss him. Then I told my cousin. He told my mom and she told my grandpa. Everything got bad after that. But still, the thought of kissin' him, or even huggin' him made me happy. So we didn't do any of that stuff but we held hands once. We're just friends though, nothin' more. I don't want any of that trouble 'cause, lemme tell you somethin'. All that talk did *not* make my family family happy. So I been talkin' to this guy who lives near me. He's *familia* too, like the way Jarod is. He tells me that at some point, I'll be able to just feel happy and say, "Fuck you!" to everyone else and then I can just be me. "Me" doesn't always make me happy 'cause everyone else is always unhappy but maybe, someday, it will.

Castes and Conversations: Kareena's Sample

The sample below is part of Kareena's narrative. Kareena was a 36-year-old woman who had spent her days working at her father-in-law's electronics shop before moving to the shelter. Her parents had lied on a dating website about their combined income, and, because of this lie, she had been able to marry into another social caste. While some argue that the caste system is no longer present in India, many social scientists continue to study its impact on contemporary society and have shown that the caste system continues to impact daily life for many across the country (e.g., George et al., 2019, 2019). Upset about this misrepresentation of income, Kareena's in-laws pressured her to provide the remaining dowry by threatening to expose her parents' deception. Her husband, feeling ashamed for not finding a "suitable" match, was physically and emotionally violent.

As a result of consistent emotional abuse, Kareena described finding relationships outside of her family as one way to provide happiness. The relationship that she fostered the most was with a maid who came to clean the house. While Kareena described multiple qualities of Madhu, the family housekeeper, that made her feel content, she focused on class as a point of bonding but also a point of contention with her family. Their relationship eventually ended in Madhu being let go, as Kareena's husband and his family felt that the relationship distracted the maid from her duties and that it could potentially expose the class disparity between their and Kareena's families.

Sample:

KAREENA: What makes me feel happy? Right now? Perhaps remembering all the good things and people in my life. There was this girl, Madhu, she was our maid. She was so easy to talk to and having her around made me happy. She was not like everyone else. She would listen. *Actually listen.* No one listens to me. It's like I'm invisible unless someone wants something. Or until my husband is mad. Then, suddenly, I'm right there. Madhu, she would just listen and talk, and, for a while, we were normal people. We were like sisters. She would clean the kitchen while I cooked, and we would just talk about her day or my day. For a while, before I came here, I still saw her around a bit, but I just thought it best not to talk to her anymore. I know it's not a normal friendship and that I shouldn't have been so close to her but, the reality is that I couldn't take any more weight from the family. It was already hard enough hearing about the dowry and how we did not pay enough. I did not need to hear about how my friendship, how her companionship, made things worse. Anyway, I suppose it doesn't matter now. I'm here. She's there. It does hurt, feeling like I lost a sister in all this. But I have myself to worry about now. Myself and this journey ahead.

A Pathology of Happiness

Tracing affects is like the light bike scene from the movie *Tron*. Tracing the narratives can illuminate personal histories and provide a space to consider other intensities that have interrupted or crossed paths with the narrator. Only, rather than tracing a single line, imagine the path as omnidirectional, stretching out across spaces and places to touch and be touched by bodies across contexts and time. Qualitative researchers across disciplines have valued stories as they are situated in, against, and through broader contexts. There are, as Clandinin (2007) argues, varying positions on whether tracing stories should stand as an act of intervention—an interruption of oppressive norms and values—or simply as a description of events. Within this text, stories not only serve as accounts of participant experiences but also, as is particular to pathology, are used to think about how one might resist, refuse, or disrupt the violence that participants articulate in their daily lives. To borrow from Sedgwick (2002), the process of touching and feeling these narratives helps interrupt the violent feedback loops that are central to hegemonic practices.

It can be challenging to cause such disruptions to oppressive norms because, as Deleuze and Guattari (1987) argue, assemblages are composed of rhizomes that have no beginning or end, no hierarchical structure, and de- and reterritorialize through multiplicity. Tracing affects within this rhizomatic terrain of an assemblage is therefore difficult. Conceivably, where one affect deterritorializes, another can form from another node within the assemblage. This is like the Hydra of Greek mythology, only the "heads" can grow in any place and without reason or provocation.

Further complicating these already complex intra-actions, Massumi (1987) explains the inherent fluidity found in what Deleuze and Guattari call *lingne de fuite,* or "lines of flight". Massumi argues that the translation of "line" is ultimately incomplete, as the term translates "not only to the act of fleeing or eluding but also to flowing, leaking, and disappearing into the distance" (p. xvi). Scholars have continued to think about the complex nature of how bodies intra-act within affects that flow within, through, and around events. This includes those like Helfenbein (2010), who describes how sociocultural norms "speak and leak" from classrooms to communities, Gums (2016), who uses poetic prose to describe the spilling of Black feminism across contexts, and Massey (2005), who argues for the fluid multiplicity of space, among others.

How, then, might one trace happiness as it moves within and erupts out from an assemblage of violence? To what end might one trace these movements? The samples above show how happiness becomes braided with other affects like sorrow or disappointment. Given these affective relations, one can argue that cartographies of happiness become more complex as they move through an assemblage. One way to describe the space these stories occupy is through what Halberstam (2005) describes as a queer temporality, or a space that allows people to believe in their futures outside of the agentic contingencies that often limit their ability to move within and across contexts.

Take José's narrative, in which, his time with Jarod can be read in several ways. One way is that he is refusing racism within his community. Yet another might be that he is calling out racism as a way to get his siblings to back off. Finally, their relationship could be understood as speaking back to heteronormative culture across questions of race. In a place predicated on white, cis-hetero values that are tied to "normal" life events (like graduation, marriage, and having children), values that are codified by straight and often racially segregated relationships nested with questions of class, José's bond with Jarod signifies a way of being, knowing, and doing that resists "conventional forms of association, belonging, and identification" (Halberstam, 2005, p. 10) . This kind of relationship is often located in a queer temporality.

Using pathology as method is one way to trace participants' stories not only as they exist independently but also as they move in comparison to what might be labeled "predictable" movements. This method speaks to personal identity, how one is identified, and how affects might continue to travel as a product of the resulting intersections. Using José's and Kareena's samples, in the next section I think through the implications of using pathology as a method to trace happiness as it moves through participants' experiences and with/in the assemblage.

Margins: Individual Agents of Happiness

In medical pathology, the term "margins" denotes healthy tissue that sits at the edges of a disease, usually a tumor. Across social sciences, the "margins" are not generally considered a positive place to those in the dominant group. However, as scholars like Goffman (1961), Spivak (1988), Cohen (1997), and Simpson (2014) have argued, while margins are often places of oppression, they are also where those who are marginalized can carve out a safe space. Greenwich Village's Hudson River waterfront in New York during the 1960s–1980s offers an example of how marginalized people have historically carved out such spaces against sociopolitical oppressions. In the 1930s, the active piers thrived with LGBTQ+ nightlife, but when the maritime industry shifted by the mid-1960s, the piers became abandoned. The vacant area only spurred more queer cruising, however, making it a popular location for gay men who were negotiating the same antiqueer norms that eventually led to the Stonewall uprising (Chauncey, 1994). Although the 1980s pier improvement project and AIDS epidemic marked a shift in pier culture, by that time the area not only served as a place for public sex, sunbathing, and drugs, but also had become a place of refuge for queer youth of color. Delaney (1988) describes how the socially disenfranchised found emotional support at the docks, despite imminent threats of violence from both the police and within dock culture itself.

Attending to José's sample above, although their space was not as formalized as the piers, he and Jarod did carve out a space that resisted racial and heteronormative segregation. While he remained unprepared to say "fuck you" to those who did not support his sexual orientation or his feelings for an African American boy living in a predominantly Mexican neighborhood, he managed to forge a space of care for Jarod and find a mentor in an older, queer neighbor. Jarod, as José notes, made him happy. His happiness marks not only emotional but physical safety, as when Jarod provided him with a knife for protection. Additionally, Jarod and José's neighbor both provided emotional refuge as, within their chosen relationships, they offered spaces where José felt heard and respected. However, happiness is not the same as a sense of belonging. While José *felt* happy with Jarod, he remained unsure that he could ever *be* happy given the broader sociocultural norms in his community, a point to which I return below.

While the piers do not offer a similar analogy for Kareena's sample, there are many historical examples of women carving out spaces to breathe against misogynistic practices and policies. Similar to José's sample, Kareena also articulated sociocultural pressure as she discussed the ideal of "normal friendships". Her inability to participate in normalized class-based relationships was also exacerbated by shame, both the shame she felt for her family's inability to pay the full dowry and the shame she experienced in finding a kindred spirit who happened to share common grounds in terms of class. However, despite this shame and consistent tension, she still thrived through her chosen relationship with Madhu, whom she described as someone who "actually listened" to her and who became a calm space amid the uncertain waters of emotional abuse. While she withstood a tremendous amount of pressure, she eventually succumbed to the growing malignancy that was class-based oppression and toxic masculinity and left her space on the margin for the shelter, a place that could be considered yet another margin.

These samples underscore the anthropological tendency to make the strange familiar and the familiar strange (Spindler & Spindler, 1982), especially for shared norms and values that are marginalizing, like those experienced by José and Kareena. The margins are often one of the last safe spaces against a tide of normalized bias, but ultimately they are also inherently negative, as oppressive norms press at and too often take over the healthy spaces created within the margins. In José's sample, he is unsure about happiness as a permanent condition because part of him desires a sense of belonging within the toxic space. The "tumor", in this case homophobia and racism, is José's golden egg—a valuable source of community and heteronormative power. He considers sacrificing his happiness for the very oppressive values that push at his own space in the margins.

In Kareena's sample, she is clear that the margin offered a temporary respite from patriarchal and class-based ideals. When the margins were pushed out and toxic ways of being, knowing, and doing impacted her space, she refused the norms by moving to the shelter. Both José's and Kareena's samples are iterations of the margins, though the samples show the difference between resistance and refusal. Within the echo chamber of oppression that forms and informs one's agentic contingencies, there are several factors at play. Although Kareena struggles with class- and gender-based social exclusion, she is able to refuse systems of oppression differently than José, who also struggles with questions of race, sexual orientation, and class. José is also a minor and therefore much less able to refuse a system within which he is legally dependent. While some people might still consider Kareena the property of her husband, her ability to move as an adult is very different than José's, who as an economically disenfranchised racial minority is further excluded through his sexual orientation.

Comparisons: What Happens at the Margins

Pathology often includes the comparison of cases. Recall, for example, how Katie's sample included several comparisons, like between the varying affects of her mother, Bryiann, and herself. To be clear, I do not use pathology as a tool to trace "normal" inroads to affects like happiness. This is because normal is often placed on a false binary with abnormalities. In the case of marginalized people and groups, this binary is too often used as a mechanism for dehumanization, as seen with the original use of anthropathology or in the way that groups like the LGBTQ+ community are often treated. Further, given the number of chosen families fostered by women, LGBTQ+ people, people of color, and other marginalized groups, the margins are far from "abnormal" (McCready, 2004). They are, however, places that are constantly and continuously threatened by broader oppressions in ways that are often normalized as business as usual. As a result, I use a pathology of happiness here to interrupt hegemonic norms rather than to engender and maintain the toxicity that created the margins.

In Kareena's sample, she outlines "normal" relationships within ideas and ideals of class that are central to the historical and contemporary iterations of the caste system in India. As scholars like Stocker (2017) and Kumar (2017) have discussed, friendships are still largely guided by class and gender-based understandings in India, even while often conceptualized as egalitarian by younger generations. In other words, class and gender are a feedback loop that means that no matter whom a person wishes to choose in their relationships, they often find themselves inescapably enmeshed with traditional norms. One can therefore think about the pathology of happiness as it travels through assemblages of violence—always already connected to oppressive norms and values but not absent of affects like hope and joy.

Happiness traveled among familiar and familial lines for Kareena as she felt comfortable with Madhu in terms of class. Her sense of belonging was ultimately tied to her ability to experience a chosen relationship with a woman with whom she shared similar experiences and histories. However, the malignancy, in this case class-based violence, manifested in terms of shame and eventually resulted in Madhu's removal from Kareena's company. Kareena was pushed to the margins, a place where she felt safe in Madhu's company, through violence. Thinking about Kareena's narrative as it compares to those shared by scholars like Stocker (2017) or Kumar (2017), the future path of her oppression might become clear. Returning to quantum entanglements, not only Kareena's connection to Madhu but also her entanglement with broader cultural values shows how the measurement of oppression forced Kareena to fall into prescribed ways of being, knowing, and doing.

Kumar (2017) argues that a large part of friendship is performance. If you visit India, for example, it is not unusual to see young men embracing each other as they walk. This performance, Kumar notes, is as significant as the deep bonds that "friendship" signifies in many Indian languages. This is not unlike Goffman's (1956) discussion of the performance of self, in which he argues that social interaction is akin to a theater performance, with each person playing a variety of roles. Goffman explains that people play roles in order to manage others' impressions of them in ways that will aid in gaining social capital. The performance of friendship in particular is often accompanied by gendered and classed understandings (Kumar, 2017), and Kareena's return to her expected role was a performance of self-preservation. Similar to Ahmed's (2010) discussion of happiness, Kareena acted out of social obligation that was connected to personal safety. Pathology might suggest an interruption of these norms, while Kareena's future desire to belong within such oppressions might continue to limit her ability to foster healthy relationships across class.

Similarly, José performed roles of straightness at home and at school that were tied to ideals of masculinity found within Latino American culture. When José's performance of a straight social identity crumbled when he was outed to his family, he found refuge in the margin. Watching happiness travel through José's narrative, it is clear that his margins were defined by cis-heteropatriarchal and raced norms and values. Yet, the performance of cis-hetero ways of being is complex across Latinx cultures (González, 2019; Moraga, 2011). As Moraga (2011) describes, queerness in communities of color often is shaped by colonized ways of being, knowing, and doing. The queer temporality that Halberstam (2005) discusses, or what Moraga argues are spaces of sustaining hope and happiness, are often double(d) in their oppression (Lather, 1991)—once by the colonization of Black and brown bodies and again by white ideals of queerness. In José's case, these ideals were further marginalized by the way that communities of color carve out space in a city, sometimes against other communities of color but always because of overarching racist policies and practices that undergird city spaces (Helfenbein, 2011; Rist, 1973; Soja, 2010). The pathology of happiness in José's sample is therefore tied to sociopolitical and historical racist and homophobic values.

Conclusion: Tracing Trajectories of Happiness

Scholars like Deleuze and Guattari (1987), Lefebvre (1974), Massey (2005), and McKittrick (2006), to name a few, have argued about the fluid, complex intra-actions that are central to the cartography of spaces and places.

Assemblages, as McKittrick (2006), Puar (2013), and Weheliye (2014) have argued, can be understood as cartographies of sociopolitical and cultural histories as they intersect with contemporary moments, movements, and bodies. Mapping an assemblage, like any other mapping project, is difficult. As Massey (2005) argues, "Space is the product of interrelations; as constituted through interactions [within]...a sphere of possibility...of multiplicity...that is always under construction" (p. 7). Assemblages, too, are always under construction within processes of deterritorialization and reterritorialization. Within these spaces that are being made and unmade, assemblages move like shuttling intensities (Gregg & Seigworth, 2010, 2010), entangling with bodies as they go.

As discussed in the previous chapter, violence is agential, meaning that it shapes how bodies act and react, or how they intra-act and construct ways of being, knowing, and doing. Assemblages of violence are particularly insidious because violence is not separate from hope, joy, and happiness. Rather, they are imbued with these affects, which become part of feedback loops of aggression across layers of violent scale. Given that violence has agency, understanding how affects move within and bubble up from assemblages is important. Tools to trace affective movements not only allow us to see their past and current movements, but also allow us a way to think about how various forms of oppression might be interrupted.

Pathology as a method offers one way to trace the fluid trajectories within an assemblage. By attending to the margins and the affects produced within those margins in relation to oppression, one can think about how these affects might be territorialized within particular intra-actions. The samples in this chapter show how happiness converges and diverges, and how it sticks between bodies and beings, within an assemblage. Like violence, happiness can be understood as agential in creating the margins and in pushing back against oppressive norms and values. The samples show how happiness, like all affects and spaces, exists in multiplicity. Each person's happiness is unique to their narrative and the intra-actions that constitute their experiences. The margins are therefore fluid spaces of multiplicity, formed and informed by broader oppressions.

My process of tracing therefore reflects the fluid nature of affects, with an understanding that one can never predict the outcome of such movements with certainty and that trying to do so can create the self-fulfilling prophecies that this method is intended to interrupt. Pathology can only look at what has transpired and use those narratives in relation to other stories to think about how future movements might be shaped and how they might shape an assemblage, be central to the affect's trajectories, or become significant to the bodies they touch along their paths.

Finally, the samples throughout this chapter also demonstrate the significance of chosen relationships in the face of daily oppressions. In documenting the rhizome of happiness, my use of pathology is meant to determine future directions. In all of the samples above, chosen relationships created inroads to happiness. Happiness can be understood to have moved through these relationships, potentially strengthening the participants' abilities to hope and feel happy within normalized systems of violence. Emerging from this method of pathology, happiness and the relationships that engender and maintain it are central to ways of being, knowing, and doing. Creating more inroads for these relationships is therefore crucial to engendering spaces from which happiness can speak and leak (Helfenbein, 2010) into violent assemblages, and to forming resistance and resilience to broader aggressions. This is especially significant for marginalized groups, for whom happiness is largely absent from the places it should most often exist—in schools and at home.

References

Agar, M. H. (1996). *The professional stranger: An informal introduction to ethnography* (2nd ed.). San Diego, CA: Academic Press.

Ahmed, S. (2010). Killing joy: Feminism and the history of happiness. *Signs, 35*(3), 571–594.

Anyon, J. (2008). Introduction: Critical social theory, educational research, and intellectual agency. In *Theory and educational research*. New York, NY: Routledge.

Barad, K. (1999). Agential realism: Feminist interventions in understanding scientific practices. In M. Biagioli (Ed.), *The science studies reader* (pp. 1–11). New York: Routledge.

Bartlett, S. J. (2005). *The first comprehensive psychology of human evil. The pathology of man: A study of human evil. Springfield.* IL, US: Charles C Thomas Publisher.

Baum, L. F., Copelman, E., & Denslow, W. W. (1939). *The wizard of Oz*. New York, NY: Doubleday.

Berlant, L. (2011). *Cruel optimism*. NC: Duke University Press.

Camus, A. (1938/1970). Nuptials. In E. C. Kennedy (Ed.), *Lyrical and critical essays* (pp. 63–106). New York, NY: Vintage Books.

Chauncey, G. (1994). *Gay New York: Gender, urban culture, and the making of the gay male world, 1890–1940*. New York, NY: Basic Books.

Clandinin, D. J. (Ed.) (2007). *The handbook of narrative inquiry: Mapping a methodology*. Thousand Oaks, CA: Sage.

Cohen, C. (1997). Punks, bulldaggers, and welfare queens: The radical potential of queer politics? *GLQ, 3* ((4), 37–65.

Cottrell, L. (2016). Joy and happiness: A simultaneous and evolutionary concept analysis. *Journal of advanced nursing, 72*(7), 1506–1517.

Delaney, S. R. (1988). *The motion of light in water: Sex and science fiction writing in the East Village*. New York, NY: Arbor House.

Deleuze, G., & Guattari, F. (1987). *A thousand plateaus: Capitalism and schizophrenia*. Minneapolis, MN: University of Minnesota Press.

Denzin, N. K., & Lincoln, Y. S. (2008). *Introduction: the discipline and practice of qualitative research*. In N. K. Denzin, & Y. S. Lincoln (Eds.), *Strategies of qualitative inquiry* (pp. 1–43). Thousand Oakes, CA: Sage Publications, Inc.
Denzin, N. K. (2017). Critical qualitative inquiry. *Qualitative Inquiry*, *23*(1), 8–16.
Dernikos, B. P., Lesko, N., McCall, S., & Niccolini, A. (Eds.) (2020). *Mapping the affective turn in education: Theory, research, pedagogy*. New York, NY: Routledge.
Dewey, J. (1916). Democracy and education. In J. Boydston (Ed.), *The middle works, 1899–1924*. Carbondale, IL: Southern Illinois University Press.
Evans-Winters, V. E., & Esposito, J. (2010). Other People's daughters: Critical Race feminism and Black Girls' education. *Educational Foundations*, *24*, 11–24.
Feltham, C. (2007). *What's wrong with us?: The anthropathology thesis*. West Sussex, UK: John Wiley & Sons Ltd.
Fosha, D., Siegel, D. J., & Solomon, M. (Eds.) (2009). *The healing power of emotion: Affective neuroscience, development & clinical practice*. New York, NY: WW Norton & Company.
Freire, P. (1992). *Pedagogy of hope: Reliving pedagogy of the oppressed*. New York, NY: Bloomsbury Academic.
Geertz, C. (1973). *The interpretations of cultures*. New York, NY: Basic Books.
George, S., Yadav, M., & Inbanathan, A. (Eds.) (2019). *Change and mobility in contemporary India: Thinking M.N. Srinivas today*. New York, NY: Routledge.
Gershon, W. S. (2015). Double binds, abuses, and a hopeless Hope: Epistemological possibilities and sensual questions for Spivak's introductory framing of an aesthetic education in the era of globalization. *Critical Literacy: Theories & Practices*, *9*(1), 4–16.
Gershon, W. S. (2017). *Curriculum and students in classrooms: Everyday urban education in an era of standardization*. Landham, MD: Lexington Books.
Giroux, H. (1997). *Pedagogy and the politics of hope: Theory, culture, and schooling: A critical reader*. New York, NY: Routledge.
Goffman, E. (1961). *Asylums: Essays on the social situation of mental patients and other inmates*. Garden City, NY: Doubleday.
González, L. C. (2019). Queerness in Latina/o/x literature. In P. Rabinowitz (Ed.), *Oxford research encyclopedia of literature*. New York, NY: Oxford University Press.
Gregg, M., & Seigworth, G. J. (Eds.) (2010). *The affect theory reader*. Durham, NC: Duke University Press.
Gums, A. P. (2016). *Spill: Scenes of black feminist fugitivity*. Durham, NC: Duke University Press.
Halberstam, J. (2005). *In a queer time and place: Transgender bodies, subcultural lives*. New York, NY: New York University Press.
Halberstam, J. (2011). *The queer art of failure*. Durham, NC: Duke University Press.
Helfenbein (2004). A radical geography: Curriculum theory, performance, and landscape. *Jct*, *20*(3), 67–75.
Helfenbein, R. J. (2010). Thinking through scale: Critical geographies and curriculum spaces. In E. Malewski (Ed.), *Curriculum studies handbook: The next moment* (pp. 304–317). New York, NY: Routledge.
Helfenbein, R. J. (2011). The urbanization of everything: Thoughts on globalization and education. In S. Tozer, B. Gallegos, & A. Henry (Eds.), *Handbook of research in social foundations of education* (pp. 319–326). New York: NY: Routledge.
Helfenbein, R. J., & Huddleston, G. (2013). Youth, space, cities: Toward the concrete. *Taboo: The Journal of Culture and Education*, *13*(1), 2.

Ibrahim, A. (2015). Body without organs: Notes on deleuze & guattari, critical race theory and the socius of anti-racism. *Journal of Multilingual and Multicultural Development*, *36*(1), 13–26.

Kumar, N. (2017). The performance of friendship in contemporary India. In C. Risseeuw, & M. van Raalte (Eds.), *Conceptualizing friendship in time and place* (pp. 229–249). Boston, MA: Brill.

Ladson-Billings, G. (1998). Just what is critical race theory and what's it doing in a nice field like education? *International journal of qualitative studies in education*, *11*(1), 7–24.

Lather, P. (1991). *Getting smart: Feminist research and pedagogies with/in the postmodern*. New York, NY: Routledge.

Lefebvre, H. (1974). *The production of space*. Hoboken, NJ: Wiley-Blackwell.

Massey, D. (2005). *For space*. Thousand Oaks, CA: Sage Publishing.

Massumi, B. (1987). Notes on the translation and acknowledgements. G. Deleuze & F. Guattari, *A thousand plateaus: Capitalism and schizophrenia* (pp. XVI–XX). Minneapolis, MN: University of Minnesota Press.

Massumi, B. (2002). *Parables for the virtual: Movement, affect, sensation*. Durham, NC: Duke University Press.

McKittrick, K. (2006). *Demonic grounds: Black women and cartographies of struggle*. Minneapolis, MN: University of Minnesota Press.

Miller, J. (2005). *The sounds of silence breaking: Women, autobiography, curriculum*. New York, NY: Peter Lang Publishing, Inc.

McCready, L. T. (2004). Some challenges facing queer youth programs in urban high schools: Racial segregation and de-normalizing whiteness. *Journal of Gay & Lesbian Issues in Education*, *1*(3), 37–51.

Moraga, C. (2011). *A Xicana codex of changing consciousness: Writings, 2000–2010*. Durham, NC: Duke University Press.

Moten, F. (2003). *In the break: The aesthetics of the black radical tradition*. Minneapolis, MN: University of Minnesota Press.

Noblit, G. W. (2004). Reinscribing critique in educational ethnography: Critical and postcritical ethnography. *Foundations for research: Methods of inquiry in education and the social sciences*, 181–201.

Ortner, S. B. (2006). *Anthropology and social theory: Culture, power and the acting subject*. Durham, NC: Duke University Press.

Pillow, W. S. (2015). Policy temporality and marked bodies: Feminist praxis amongst the ruins. *Critical Studies in Education*, *56*(1), 55–70.

Puar, J. (2013). 'I would rather be a cyborg than a goddess': Intersectionality, assemblage, and affective politics. *Meritum*, *8*(2), 371–390.

Rist, R. (1973). *The urban school: A factory for failure*. Cambridge, MA: MIT Press.

Robbins, S. (2010). *Robbins and Cotran pathologic basis of disease (8th edition)*. Philadelphia, PA: Saunders.

Rosaldo, R. (1993). *Culture & truth: The remaking of social analysis*. Boston, MA: Beacon Press.

Schaefer, D. (2013). The promise of affect: The politics of the event in Ahmed's *The Promise of Happiness and Berlant's Cruel Optimism. Theory & Event*, *16*(2), https://www.muse.jhu.edu/article/509908.

Sedgwick, E. K. (2002). *Touching feeling: Affect, pedagogy, performance*. Durham, NC: Duke University Press.

Simpson, A. (2014). *Mohawk interruptus: Political life across the borders of settler states*. Durham, NC: Duke University Press.

Skattebol, J. & Hayes, D. (2016). Cracking with affect: relationality in young people's movements in and out of mainstream schooling, Critical Studies in Education, 57(1), 6–20

Snaza, N., Sonu, D., Truman, S. E., & Zaliwska, Z. (Eds.) (2016). *Pedagogical matters: New materialisms and curriculum studies*. New York, NY: Peter Lang Publishing.

Snowber, C. N. (2013). Dancing a curriculum of hope: Cultivating passion as an embodied inquiry. *Journal of Curriculum Theorizing*, *28*(2).

Soja, E. W. (2010). *Seeking spatial justice*. Minneapolis, MN: University of Minnesota Press.

Spindler, G., & Spindler, L. (1982). Roger harker and schonhausen: From the familiar to the strange and back again. In G. Spindler (Ed.), *Doing the ethnography of schooling*. New York, NY: Holt, Rinehart and Winston.

Spivak, G. C. (1988). Can the subaltern speak. In C. Nelson, & L. Grossberg (Eds.), *Marxism and the interpretation of culture* (pp. 271–313). Basingstoke, UK: Macmillan Education.

Stewart, P. (2007a). Who is kin? Family definition and African American families. *Journal of Human Behavior in the Social Environment*, *15*(2–3), 163–181.

Stewart, K. (2007b). *Ordinary affects*. Durham, NC: Duke University Press.

Stocker, S. (2017). *Caste and equality: Friendship patterns among young academics in urban India*. Germany: Transcript-Verlag.

Tapper, M. (1997). An 'Anthropathology'of the 'American Negro': Anthropology, genetics, and the New racial science, 1940–1952. *Social history of medicine*, *10*(2), 263–289.

Tomkins, S. S. (1962). *Affect imagery consciousness: Volume I: The positive affects*. Netherlands: Springer

Tseng, M., Hu, Y., Han, W-W., & Bergen, B. (2005). "Searching for happiness" or "full of Joy"? *Berkeley Linguistics Society*, *31*(1), 359–370.

Tsing, A.L. (2005). *Friction: An ethnography on global connection.* Princeton, NJ: Princeton University Press.

Weheliye, A. G. (2014). *Habeas viscus: Racializing assemblages, biopolitics, and black feminist theories of the human*. Durham, NC: Duke University Press.

Werner, R. A., & Benjamin, B. E. (2002). *A massage therapist's guide to pathology*. Lippincott Williams & Wilkins.

Wild, E. (2005). Deja vu in neurology. *Journal of neurology*, *252*(1), 1–7.

Wrigley, T., Lingard, B., & Thomson, P. (2012). Pedagogies of transformation: Keeping hope alive in troubled times. *Critical Studies in Education*, *53*(1), 95–108.

Yin, R. K. (1981). The case study crisis: Some answers. *Administrative science quarterly*, *26*(1), 58–65.

4

CURRICULA OF VIOLENCE

Learning In And Through The Assemblage

In the fall of 2017, I began my work with the group that would later call themselves the "Blue Dragons". While the goal of this project was to think with students about sexual orientation and gender identity in all of their multiple forms and expressions, part of my work as an ethnographer was also to understand the context of the school, the city, and the students' lives. In order to think about these complex, interrelated contexts, students agreed to complete mapping projects to help me gain a better understanding of the school. Using contemporary debates over mapping in critical geography and schools (e.g., Gershon, 2013; Helfenbein, 2004; Kitchin, Perkins, & Dodge, 2009; Nespor, 1997; Powell, 2010), I requested that students complete two stages of mapping: one auditory and one visual.

During the visual mapping portion, students recorded their time at school by drawing, painting, or constructing maps from modeling clay. Their creations ranged from general reference maps that resembled traditional street maps to narrative maps that combined symbols and pictures. I noticed divisions along lines of gender identity and sexual orientation that were intriguing. For example, students who identified as cisgender girls or femme lesbians and boys who identified as gay tended to draw maps that were more narrative in nature. In contrast, most students who identified as cis-boys and some butch lesbians drew reference maps. This division was not true for every student, and there were combinations of these map styles that emerged as students worked near each other. However, there was a clear correlation between students' identities and the mapping style that they preferred. While the outcome and analysis of these maps is not the point I focus on here, it is worth noting the diversity of findings that were produced from this first part of the project as they seemed to correlate with questions of identity and orientation.

Students completed the auditory mapping in two rounds. First, I asked them to carry an audio recorder to tell me about places that were significant to them and, should they feel comfortable, to share why that particular place was important. Of the 14 students in the group, two recorded classrooms as significant places in the school. Other locations included the cafeteria, the schoolyard, the bathroom, and the administrative assistant's area in the main office. The most recorded place was the school's chicken coop. A few weeks later, I asked them to repeat this activity with a slight variation: I wanted them to record their impressions of the places where they felt they learned the most on the school grounds. In this second iteration, only four students recorded classrooms. The chicken coop continued to be an overwhelmingly popular spot and was accompanied with several narratives about the chickens and their lives.

There are several reasons that might explain why the chicken coop was so regularly recorded, including students' explanations that time there was considered a special privilege. Many students in the group complained that they were excluded from time near the coop compared to their peers, who they perceived as being granted more time to observe the chickens than students who participated in the Blue Dragons. There are also several possible reasons for students not acknowledging classrooms as either personally significant or as places for learning. This includes being apprehensive about recording a teacher's space, especially if a class was in session. However, most students did not even enter the building when they recorded the places they felt held special significance to their time at New Horizons. Even fewer students entered the building to record spaces they felt were significant to learning. Additionally, many students asked to take the recorders home to continue recording places so that I could, as one student described, "hear [her] home and the places that *really* mattered".

At the same time the students in the Blue Dragons group were mapping their school, teacher candidates at the institution where I was completing my postdoctoral research fellowship were working on a soundscape project in my Education 101 course. Following prominent dialogues on sound and how to think critically about what is learned through listening (e.g., Feld & Brenneis, 2004; Gershon, 2017b; Oliveros, 2005; Pinch & Bijsterveld, 2012, 2012; Schafer, 1969), students were asked to create a soundscape that answered the question, "What does learning sound like?" Accompanying the soundscape, they were to write a one-page rationale using course readings that explained which sounds were included, which were excluded, and which were foregrounded in the recording. These education students overwhelmingly recorded the sounds of children in school, and, of all the places in schools, they generally recorded classrooms. Of the 25 students in my class, only three included sounds from outside of schools as places they identified as significant for learning.

There are many reasons why these education students might have chosen to record schools over, for example, the subway. While discussing their process in class, undergraduates cited that schools were more accessible to them than many places because of their field placements for another course in their program. They also overwhelmingly explained that they had not considered places outside of schools as educational spaces. Regardless of the reasons, the stark difference between what middle school students of color described as places to learn and the educational spaces identified by overwhelmingly white teacher candidates became for me what Agar (1996) calls a rich point, or an unexpected turn of events that "signals a gap between worlds" (p. 31).

As a woman of color who, like the students at New Horizons, found places outside of school to be far more educational than classrooms, this disconnect was not shocking to me, especially as someone who attended a predominantly white institutions for my graduate schooling. This disparity is certainly not new and is one that educational theorists have continuously explored, repeatedly noting the misconception that classrooms are the sole authority of teaching and learning (e.g., Cooper, 1892; Metz, 1978; Oakes, 1985; Spencer, 1859; Varenne & McDermott, 1998). However, there was something about that moment that affectively stuck with me as I listened to hours of recordings at the chicken coop and then hours more of teacher candidates recording teachers performing the act and art of teaching. So, to echo the question of my own assignment: What does learning sound like? Where does it take place? How do we determine when someone has learned or to what extent they have learned?

This chapter reflects on these questions as they relate to a curriculum of violence. By a "curriculum of violence", I specifically refer to the way that violence is normalized through assemblages that are entanglements of and across educational contexts, both inside and outside of schools. While this chapter attends primarily to schools and systems of schooling, people learn across spaces and places in many ways. Most parents or guardians can tell you that a toddler understands gravity well before she learns about it at school. Similarly, learning is not always intentional. A child does not mean to learn about gravity when she falls off the couch any more than most people mean to learn about a broken heart when entering a first relationship. But just as some lessons are unintentional, others are calculated, like a parent explicitly teaching hate or, echoing José's narrative in the last chapter, aspects of masculinity and sexuality. These lessons, like many forms of education, do not need the formality of a schoolroom (Myers, 2019) to legitimize their impact.

Just as learning happens in many (un)intended places, multiplicities of violence exist across contexts, histories, politics, cultures, and bodies in ways that are intraconnected through an assemblage. Understanding violence as an assemblage means attending to the idea that every act of violence is predicated on and exists in relation to past, present, and future iterations of aggression. The intra-actions of bodies, both human and nonhuman (Barad, 2007), mean that violence is constantly and consistently de- and reterritorialized in and through an assemblage. The previous chapters focused on what assemblages are, how affects entangle within them, and how we might trace the trajectories of affects within such assemblages. In this chapter, I attend to violence as a curriculum in order to explicate how we learn, engender, and maintain violence in ways that reify aggression as business as usual.

Dialogues focused on the multiple iterations of violence and their sociopolitical histories have been central across antioppressive traditions, including the African American intellectual tradition (e.g., Cooper, 1892; DuBois, 1903; Woodson, 1933), queer theory (e.g., Halberstam, 1993; Haritaworn et al., 2014, 2014), and indigenous scholarship (e.g., Atkinson, 2012; Simpson, 2014), to name a few. Violence has been theorized through the interrelated forms of symbolic (e.g., Bourdieu & Passeron, 1970; Burawoy, 2019), political (e.g., Foucault, 1978; Mbembe, 2019), and cultural (e.g., Bachman, 2019, 2019; Fanon, 1961) aggressions. Building on these complicated conversations, I pay attention to and emphasize curriculum—which is learned across contexts, from classroom to community.

The reciprocal relationship between schooling and communities with respect to normalized violence is a well-trodden subject. Woodson (1933) argued that there "would be no lynching if it first did not start in the schoolroom" (p. 3), scholars like Fanon (1961) have discussed the colonizing impact of schooling on one's ontoepistemology, and others like Gilbert (2014) and Sedgwick (1993) have discussed the social-emotional damage of schools on queer bodies that speaks to community norms and values. In sum, schools do not exist in a vacuum, and the violence that is learned in schools recurs in and through communities and, in turn, impacts schools. Schooling is therefore central to the resonances and reverberations (Gershon, 2017) found in the echo chamber of aggressions within an assemblage.

This chapter begins with schooling but acknowledges two important points. First, as discussed above, schools should not be understood as the sole authority on education. As indigenous scholars have noted, the idea of an "official" or formal curriculum (Apple, 1993) often leaves out ways of being, knowing, and doing that are central to non-Western cultures (Smith et al., 2018, 2018; Wang, 2014). The narrow and politically based curriculum presented through public schools in Western nations is designed to promote and maintain cultural genocide (Crichlow, 2002). The colonized intellectual (Fanon, 1961) normalizes Western values not only through the curriculum but also through a structure of schooling that narrowly conceptualizes how and where learning happens. Despite the need for critical schooling and pedagogy, as argued for by theorists like Freire (1970), Giroux (2018), and Woodson (1933), among others, teachers remain overly concerned about classroom management and formal curriculum rather than with dismantling neoliberal education by, for example, incorporating decolonial models of schooling.

Second, because schools in places like the United States are predicated on the ideas and ideals of eugenics (Watkins, 2001; Winfield, 2007), not only are they narrow in their conceptualizations of what learning looks like, who should hold power in the classroom, and where learning should happen, they also continue to be overwhelmingly marginalizing spaces. As I argue below, marginalization is not only an intended consequence of schools, but such violence resonates so deeply within the structure of schooling that it is difficult, if not impossible, to detangle the knots of aggression in schools (Nespor, 1997; Varenne & McDermott, 1998).

Finally, because of the vastly different contexts and purposes of the two studies described throughout this book, I use them as separate yet entwined cases that provide different lenses through which to understand the contours of a curriculum of violence. I spent my time at New Horizons thinking with middle school youth of color about queerness, living and learning in an urban context, and middle school experiences. Although students regularly shared about their home lives, my primary focus was on how those experiences spilled into the school. In my work at Aasha Women's Shelter, I aimed to think with women who were survivors of intimate partner violence about how Indian culture and their home contexts normalized violence against women and girls. While a few participants mentioned their own or their children's experiences at school, the study was predominately focused on broader sociocultural and political lessons that normalize violence, lessons that leak reciprocally between schools and communities. Because of the British influence on Indian schools, these do share some common practices with other Western nations like the United States, such as using schooling as a site to normalize Western values (Kingdon, 2020; Visweswaran, 2010). However, rather than treating these studies as a direct comparison, I use them to think about the various parts of a curriculum of violence: what is learned, how it is maintained, and the sociopolitical implications of such a curriculum, especially for marginalized people and groups.

Capitals of Shame and Schooling

ELENA: Dr. Wozolek, I can't come to the museum. Principal Stein said I would break the rule.

WOZOLEK: Which rule?

ELENA: The first rule of being here at New Horizons.

WOZOLEK: And what would that be?
ELENA: Well…. (looks around)
DONTÉ: (Looking at Elena) You want me to tell her the rule? How does she not know the rule? Should we tell her the rule?
WOZOLEK: Okay… What rule, people? Donté, José, Marissa, and Elena (together): Don't embarrass Mr. Stein!
WOZOLEK: What? All in unison, this time louder: DON'T EMBARRASS MR. STEIN!
ELENA: How do you not know that, Dr. Wozolek. It's, like, the only rule of this place.
WOZOLEK: And what does it mean to embarrass Principal Stein?
ELENA: You have to act nice.
MARISSA: And don't be too loud!
DONTÉ: And don't be a fool.
JOSÉ: And… ummm… What is that again? Oh yeah, just don't be like yourself. Be as white as possible. Act. White.
ELENA: What?
JOSE: Act white! (laughing)
MARISSA: Hey! I'm white!
JOSÉ: Not like you, girl. Like, you know, white and straight. And then more white, again. You embarrass people just by those lice you were bringin' to school.
MARISSA: (punches José lightly in the arm) Okay, whatever. Anyway, that's why my hair is cut off so just calm down. Not like I could help it…

Shame is a common mechanism of control in schools. It has been used to criminalize and dehumanize students (DuBois, 1903; Morris, 2016). It has been used to maintain the status quo and marginalize targeted students and groups (Farley, 2018; Gilbert, 2014), and it has been weaponized against teachers and parents (hooks, 2003; Kozol, 2005). Given the pervasive nature of shame across systems of schooling, it is worth returning to the idea of capitals of shame from Chapter 2. Capitals of shame describe how a person or group uses shaming as a tool to lessen others' agency, or to render them as incompetent or impotent as possible. These capitals of shame are so entrenched in systems of schooling that they have become normalized in schools and their broader communities. For example, dividing lines of students by gender can be shaming for genderqueer students, learning about Black history only through slavery or civil rights shames Black ontologies into narrow understandings, and round-robin reading practices shame students whose strength does not happen to be reading aloud. Because the purpose of shame is not to gain shame but rather to be able to wield it against others (Sedgwick, 2003) to gain social capital, bodies become entangled with recursions of shame. Schools therefore offer fertile ground on which to think about the norms of shaming, because shame exists across the curriculum and the intra-actions that are normalized in school.

As curriculum scholars have noted, most people involved in schooling—from students to teachers, from administrators to board of education members—understand "the curriculum" through a rather narrow lens. This is often conceptualized as a planned sequence of instruction constructed by teachers, school districts, or, in a neoliberal era of education, national goals and guidelines (e.g., Barnes & Gonzalez, 2015; Wiggins & McTighe, 2005). Although thinking through what, when, and how to teach is an important endeavor, these are not the sum total of the curriculum. Rather than holding onto this limited understanding, here I align with the field of curriculum studies, which theorizes that learning in school is driven not by *the* curriculum but rather through four forms of curriculum: formal or official, null, hidden, and enacted. I describe these in detail in the next section.

These forms of curriculum are impossibly entwined. They are responsible for shaping the sociocultural norms of a school as well as how and what students learn as a result of their time there. This distinction is important because the traditional lens only attends to formal lessons and essentially negates the responsibility of those within a schooling system for reifying broader norms and values. For example, in the United States, according to the constitution, there is supposed to be a separation between church and state. This division extends to public schools and, as such, the curriculum is often understood as sanitized from broader sociopolitical dialogues. But as Apple (1993) argues, "the curriculum is never simply a neutral assemblage of knowledge" (p. 223). Here, I pay attention to how that assemblage of knowledge is central to an assemblage of violence. Understanding how the forms of curriculum are always at play helps to unpack the messiness of what and how youth learn and, in turn, how broader norms and values are established. The following sections break down the four forms of curriculum to show how they are entangled with shame and violence at school.

The Four Forms of Curriculum

DAVONTÉ: What do I learn in school? I don't know. Whatever they want me to learn.

Whatever is going to help me get into a better high school. It's not shit that I *need* to learn but it's definitely stuff that they *want* me to learn.

In 1918, John Franklin Bobbitt wrote *The Curriculum,* one of the first recognized texts in the United States to think about school curriculum in this fashion. Bobbitt, a prominent eugenicist (Watkins, 2001; Winfield, 2007), used his platform as an educational theorist at the University of Chicago to outline a theory of curriculum that was aligned with racist, sexist, and classist ideologies (Winfield, 2007). From his work, other educational theorists like Edward Thorndike, G. Stanley Hall, and W.W. Charters began to construct a curriculum theory that was based in eugenicist norms and values. While contemporary U.S. education rests on this foundation built with classist, ableist, racist, and sexist ideas in mind, there were other prominent theorists like John Dewey (1938), whose discussion of education as experiential has deeply shaped teacher education today. Similarly, Joseph Schwab's (1969) work focused on the deliberative nature of an intended curriculum, influencing how the formal curriculum is understood as always in action.

Unfortunately, contemporary teacher education programs tend to focus on the framework proposed by eugenicists and accepted by the U.S. culture that was permissive of such ideals. While there is often attention to the work of scholars like Dewey, who emphasize experience in education, the voices and scholarly contributions of many African American scholars continue to be left out; this includes the work of Anna Julia Cooper (1892), Mary McLeod Bethune (1938), W.E.B. DuBois (1903), and Carter G. Woodson (1933). These educators not only discussed the formal curriculum but, in places like the *Journal of African American History* (founded in 1916 by Woodson) or the *Journal of Negro Education* (founded in 1932 by Charles Henry Thompson), they regularly published on how to build a curriculum that foregrounded Black voices, experiences, and perspectives. Take, for example, Bethune's *Clarifying the Vision with the Facts*, in which she argues for a countercurriculum by stating: when "they read of Columbus, we must introduce the Africans who touched the shores of America before Europeans emerged from savagery...with Edison we give them Jan Matzeliger; with John Dewey, we place Booker T. Washington... as we tell this story, our pride in racial achievement grows" (pp. 12–13). As Bethune argues, a curriculum of Black achievement has already been written. It is not a curriculum of the margins, though it is marginalized. But it is written as a formal curriculum.

The "formal", or what Michael Apple (1993) calls "official", curriculum is what people tend to envision when they picture *the* curriculum. It is what parents wrestled with at home as schools shifted to online learning during the COVID-19 outbreak, and it is what teachers grapple with in schools every year. However, the formal curriculum is more than just a sequence of study. It is an intended course of learning that was built on white, cis-hetero, and patriarchal understandings. It was written to recognize voices aligned with that foundation, and it was designed to silence others. It is a space that eugenicists designed to intentionally silence those that they felt "sprung from worm-eaten stock who [will]...remain to the end stupidly unresponsive to the delicate growth factors of education and...[should]

be banished to raise the race to higher altitudes" (Bobbitt, 1909, p. 385). Returning to Apple (1993), the formal curriculum, whether acknowledged or not, is always tied to sociopolitical norms and values.

The opposite side of the formal curriculum is the null curriculum, or what is explicitly not taught (Eisner, 1985). Each semester, I repeat an activity that I learned during my master's program from one of my professors, Walter Gershon (see also Gershon, 2017b). I ask students to yell out all the names of white male scientists that they can in 30 seconds. After listening to the usual list—Einstein, Newton, Darwin, Hawking, Watson, and Crick—I ask students to name all the white female scientists that they can. There is generally silence before someone proudly yells, "Marie Curie!" This usually elicits some chuckles from students that are met with more silence. I then ask students to yell out all the male scientists of color that they can. Sometimes the silence is broken by someone shouting, "George Washington Carver?" Finally, I ask students to name female scientists of color. Aside from "those women who were in the Hidden Figures movie", there are rarely any contributions. When I ask for the names of "those women", there is generally no answer. When I then ask what would happen if I asked students to repeat the activity with the names of LGBTQ+ scientists, I am generally met with stunned looks that beg me not to repeat the activity.

In addition to the absent histories and voices of marginalized people and groups, the null curriculum can also be understood as an absence of particular skills that are marginalized from formal curriculum. In the current standardized era, this includes critical thinking skills (Milner, 2017), technical skills that are generally left to vocational schools (Levesque et al., 2000; Lytvyn et al., 2020), and skills generally relegated to "the arts" (Roy et al., 2019).

The null curriculum is constructed not only of these absences but also by voices being pushed out of the formal curriculum. Returning to my discussion of pathology from the previous chapter, the null curriculum is a margin, the healthy tissue that surrounds a malignancy. By "the margins", I refer not only to those who are marginalized but also to the spaces and places that are disregarded and silenced by dominant norms and values. However, unlike the margins that José or Kareena occupied, it is difficult to create such places when the space itself is created by an absence of ideas. Students in my courses each semester are unaware of the absences created through the null curriculum; it is difficult to occupy a space that you are unaware exists.

When Davonté says that he is not learning what he "needs" to learn, but rather what "they want him to learn", he is identifying a tension that exists for many students of color, especially for queer youth of color (McCready, 2004). This tension is formed between threads of a curriculum that attend to African American sociohistorical roots as they are woven into a predominately white curriculum (Watkins, 2001) that is generally absent of all queer voices. One reaction to the null curriculum is offered by scholars like Gloria Ladson-Billings (1998), Geneva Gay (2002), and Django Paris (2012), who have written about the necessity to include culturally relevant, responsive, and/or sustaining pedagogies in schools. While the frameworks of their arguments vary, the central tenets are the same: the curriculum should be inclusive of voices resonant to the context of the community in ways that encourage critical dialogue about dominant norms and values.

While I was present at New Horizons, the teachers were actively working to incorporate African American voices across disciplines. Recalling Katie's sample from the previous chapter, she was excited that Bryiann could attend a Langston Hughes assembly to showcase student work. These kinds of events, which displayed the school's commitment to incorporating Black voices and experiences, were not unusual. However, while teachers worked to incorporate voices normally relegated to the null curriculum into the formal curriculum, the broader framework of the formal curriculum remained largely traditional in the ideals that it espoused and the ways it was implemented. This meant that eugenicist norms—like testing models, tracking, and voices that were still primarily from the white cis-hetero patriarchy—remained a salient part of students' experiences with the formal curriculum. In other words, even what could be perceived as a "counter (formal) curriculum" still created margins that were ultimately harmful to the students at the school.

The null and formal curricula are central to the classroom but, as educational anthropologists and sociologists have noted, learning happens across school contexts, from the classroom to the corridors (Metz, 1978), from the bus line to the cafeteria (Tatum, 2003). The hidden curriculum (Anyon, 2000; Giroux & Penna, 1983) represents the lessons in which students, teachers, and administrators participate but that are not necessarily intended; they are ultimately hidden to those engendering and maintaining the norms and values being taught. The hidden curriculum is, as Anyon (2000) argues, a form of social reproduction. At New Horizons, an example of the hidden curriculum is how shame from the principal became woven into daily ways of being, knowing, and doing at the school. While the expectation was to not "embarrass the principal", many other adults in the building began to point out student behavior as "embarrassing". Actions often labeled as "embarrassing" included being "too loud" in talk or laughter, any actions that might be considered "gay", or any times traditional gender roles were reversed. As one student said, "If you don't want embarrass people here, girls have to be really girls and boys really need to act like men".

The hidden, formal, and null curricula are always in action through what is called the enacted curriculum (Page, 1991). Schwab (1961) argues that the curriculum is dynamic, constantly being made through actions. Several educational theorists have focused on the teacher-to-student processes that enact a curriculum (e.g., Erickson, 1982; Gershon, 2017a; Mehan, 1979); these relations are exceptionally significant as they allow scholars to unpack how power is territorialized and deterritorialized in schools through the actions of students, teachers, parents, and administrators. I expand on this conceptualization of actions and bodies, following Barad (2007) in order to attend to the intra-actions that occur between all bodies in and around schools, both human and nonhuman.

Think about the geography of schools. Scholars have noted that schools tend to hide students with dis/abilities in classrooms that are difficult for visitors to openly observe (e.g., Abend, 2001; Gottfried et al., 2019; Page, 1991). The Department of Education in the United States currently boasts that about 35% of students with dis/abilities spend nearly 80% of their academic day with their peers in "general" classes (National Center for Education Statistics, 2019). However, students and teachers have noted that, despite this inclusion, buildings designed to hide dis/abled students away teach both students in general education tracks and those who are dis/abled to devalue the physical and neurological diversities that are not aligned with the prescribed "norm" (Valente, 2011; Yoon, 2019).

At New Horizons, the building space was limited, so groups like the Blue Dragons met in a building adjacent to the school. The building was primarily used to house the school nurse, a school psychologist, and a therapy program for youth that were facing incarceration if they had one more infraction. Prior to the establishment of other clubs, which were also eventually housed in a small room in this second building that lacked air-conditioning or proper heating, the Blue Dragons were relegated to the cafeteria. Students in the group felt that the cafeteria space was "much cooler" because it was a place that "everyone went". Tucked away above the nurses station and the juvenile diversion space, next to a psychologist's office that students noted they were only called to "when they were in 'real' trouble", Blue Dragon students explained that this second location felt "cool because they got to leave the school building" but also ultimately like "a punishment because [they] lost cafeteria privileges for the group". While there are always many explanations for why a group like the Blue Dragons could not have space in the school building, or why a classroom for students with dis/abilities is located away from the main hallways, the outcome is the same—the involved students feel alienated and other students normalize this alienation and shame.

The normalization of shaming and dehumanization therefore happens across all forms of curriculum. From the formal curriculum built on eugenicist ideologies to the null curriculum constructed by silences and absences, students like those in the Blue Dragons are shamed to the margins and, mirroring the null curriculum, are silenced among teachers and their peers.

Given the entangled nature of shame and the multiple forms of curriculum, acts of shaming found in the hidden curriculum were also imminently present through the enacted curriculum. When the principal, Mr. Stein, iterated the universal rule to not "embarrass him", this rule hit the bodies and beings of youth at New Horizons in several significant ways. First, being a predominantly African American and Latinx school, even when voices that were resonant with their home communities were added as a supplement to the formal curriculum, this inclusion was not enough to enact a sense of value among students nor, arguably, among teachers. The lessons enacted through the formal curriculum therefore carried a degree of shame, which came with placating a community but not working to fully incorporate their voices into the curriculum.

Second, as faculty embraced the principal's call for students to not be an "embarrassment", the hidden and enacted curricula became entangled with a sense of shame throughout students' daily intra-actions with faculty in general. Finally, as students intra-acted with each other, teachers, and the physical building, normalized understandings about who could occupy which spaces and under which conditions brought further shame to queer youth of color, including several in the Blue Dragons. It is no wonder that students predominately filmed places on the outside of the school like the chicken coop rather than noting classrooms, corridors, or any other indoor place as spaces of learning.

Within the spaces of learning found in a curriculum of violence, messages of white, cis, hetero, and patriarchal ideas and ideals become normalized. Queer people and groups often live and learn in bloom spaces of oppression meant to further marginalize their ways of being, knowing, and doing. While all students experienced varying degrees of shaming at New Horizons, students in the Blue Dragons group often articulated feeling "tired" by the constant onslaught of negative messages around queerness. As I have theorized elsewhere, this exhaustion can be understood as queer battle fatigue (Wozolek et al., 2015), a response to the consistent homo- and transphobias that are not only present but also taught through systems of schooling and their imbrications with broader communities. Perhaps more damaging, queer battle fatigue often leads to feelings of anxiety, depression, and, for some, the desire to self-harm. In other words, fatigue is often part of the conduit of what I have named the school-to-coffin pipeline for queer and questioning youth (Wozolek et al., 2017). Students at New Horizons faced both queer battle fatigue and the school-to-coffin pipeline as they resisted the broader curriculum of violence. While the students did not disclose suicidal thoughts, they did explain how the tensions between queerness and cis-hetero normative values made them feel exhausted. As Dominique noted, "It's just one more thing on top of everything else.

Sometimes that makes me feel tired". Given their budding identities, it is no wonder that as queer youth and queer youth of color continue in systems of schooling, their fatigue all too often turns to thoughts of self-harm.

The curriculum of violence is therefore a product of the nested and layered nature of normalized aggressions found across the formal, null, hidden, and enacted curricula. Much as DuBois (1903) posed the question, "How does it feel to be a problem?", it is worth asking how schools continually territorialize the idea of a "problem" through a curriculum of violence. At New Horizons, students were a "problem" when their ways of being, knowing, and doing were rendered as an "embarrassment". The curriculum of violence is therefore not just about shame, it is about normalizing dehumanization. When Davonté explained that he learned what they "wanted him to learn", he was learning through violence and against his ontoepistemology.

Across the United States, there is an ongoing rise in murders of LGBTQ+ people and related hate crimes (Dashlow, 2018). Under the Trump administration, hate crimes based on race and sexual orientation have surged to a 16-year high, accounting for a combined total of over 75% of all bias crimes committed in the United States (Department of Justice, 2018). For queer youth of color, represented in both "queer" and "racially motivated" hate crimes, the curriculum of violence that they experienced and that was also internalized by their peers will impact them at school and in their communities—many times in physically and emotionally violent ways. Returning to Woodson's statement about lynching and the schoolroom, is it any wonder that when queer youth of color are seen as a "problem" or an "embarrassment", they are subjected to what have been dubbed "modern lynchings" (Thomas, 2019) across the United States?

Violent Frictions and Forces: Community Curriculum of Violence

DIYA: I don't know when or where I learned that it was okay to let my husband touch me when I didn't want it. The lesson was in my mind before I was even married. I'd just let it happen, when he wanted. One time I had a major surgery and, less than a day after, he wanted to be with me. What else could I do? I just let him do what he wanted, kept quiet, and just watched the clock and listened to the wind outside. Sometimes that helps, to focus on something, to not move, to just breathe. I know it's not rape if you don't say no, right? But that night was so painful. I could feel my stitches stretching. I just focused on my breath. No one teaches you these things before your wedding night. You are alone to figure them out. Then again, no one teaches you to endure these moments with your husband but, somehow, you already learned they were supposed to be acceptable.

When the novel coronavirus became a global pandemic in 2020, lockdowns became mandatory across the globe. Social media was soon inundated with stories of hope, like people singing to each other from their separate apartments or wildlife taking over abandoned spaces. There were also stories of sorrow, like posts that shared the deaths of loved ones or news outlets that reported mass graves being dug around the world. While every country dealt with lockdown, quarantines, and social distancing differently, one narrative was consistent across the globe: the spike in domestic violence. In Spain, domestic violence hotline calls, emails, and police reports increased by nearly 700% when the lockdown began in March 2020 (Ministerio de Igualdad, 2020). In the United Kingdom, the largest domestic violence relief charity, Refuge, noted a 150% increase in hits to their website and a 25% increase in calls to their hotline (Refuge, 2020). Similar reports came out of the United States (Campbell, 2020), France (Febvre, 2020), and Canada (Bogart, 2020), among other nations around the globe.

India was no different. Most news outlets, including the *Tribune Times*, *Hindustan Times*, *The Diplomat*, and *The Times of India*, reported a sharp rise in domestic violence cases when the Prime Minister of India, Narendra Modi, put the country on strict lockdown procedures. In the first few weeks of lockdown, calls poured into helplines, with both women and children describing increased abuse. One reporter from *The Hindustan Times* wrote that by the second week in April, over 2,500 women and children had called the Emergency Response Support System in Delhi alone (Ratnam, 2020). Adding to the crisis, shelters across the country were forced to close due to health concerns. One survivor told a hotline volunteer, "*Ekhon phone kora-o mushkil-bhoy lag*" (Even calling has become difficult now—I feel afraid (Chakraborty, 2020, para. 3)). I tried several times during the COVID-19 pandemic to reach Aasha Women's Shelter, but no one answered my calls and no machine was available to leave a message. Chakraborty (2020) notes that when shelters like Aasha closed due to health concerns over COVID-19, there was little to no way to know where women were scattered, whether it was back home, to the streets, or with friends or family.

The rising number of people worldwide who have become victims of intimate partner violence as a result of the pandemic is incredible. As I write, there is no end to COVID-19 in sight, with no vaccine and no standard treatment procedures, particularly as the virus continues to mutate. This means that there is also no foreseeable end to the rise in violence, especially as future surges of the virus might necessitate additional lockdowns to contain the spread. Additionally, treating cases of violence as they are reported is like performing triage in a battle unreachable to crisis centers and hospitals, as those who respond only see the damage left by violent intra-actions.

Witnessing the aftermath of aggression is like hearing reverberations of violence but seeing what made the sound and not being able to interrupt the motion before pressure is applied to create friction or force. Recalling the discussion of friction and force from the introduction to this book, sounds are made by one or both of these actions. As Tsing (2005) argues, friction shapes bodies through their relationships. Friction and force also cause movement. Remember, too, that both the tire and the road are worn down by friction, though there are factors—like agentic contingencies—that shape which bodies might wear down more quickly or more prominently. Regardless of the wear, both bodies are shaped, and movement happens.

The women at Aasha, both pre-COVID-19 and during the pandemic, were victims of violent friction and force applied by their husbands and families. This was a directed pressure, one meant to form and inform a woman's way of being, knowing, and doing to fit the context of her husband's house. However, this pressure is not just applied by a woman's family. She also sustains the pressure of sociocultural and political friction and force. When Diya explained that she doesn't know when she learned to allow her husband to "touch [her] when [she] doesn't want it", she was not just referring to pressure from her husband, she was also responding to broader norms and values with which she is entangled on the quantum level.

In Chapter 1, using quantum entanglement, I argued that a person can entangle so deeply with sociocultural norms and values that their ways of being, knowing, and doing are shifted as culture is "measured". Here, I focus on what is learned during those entanglements, approaching them as a form of curriculum. When curriculum theorists and others discuss the curriculum, they understandably tend to focus on schools. However, because the lessons within schools are reciprocal with their surrounding communities, the four forms of curriculum can all be understood as occurring across contexts.

There are several ways to understand curricula across school and community contexts. Theorists have discussed this as public pedagogy (Giroux, 2000; Sandlin et al., 2010, 2010), through decolonial frameworks (Kovach, 2013; Patel, 2015; Sabzalian, 2019), or within the scope of culturally sustaining pedagogy that has strong community connections (Milner, 2011; Paris, 2012). Each of these frameworks is significant in that they recenter conceptualizations of education and spaces of learning beyond the colonized ideal of the classroom (Smith et al., 2018, 2018). Though I acknowledge the strengths of these pedagogies, here I will approach the assemblage itself as inherently pedagogical.

As several educational theorists have argued (e.g., Ellsworth, 2005; Green, 1978; Lusted, 1986; Massumi, 2002), there is a reciprocal relationship that territorializes and deterritorializes ontoepistemologies within the sensations and movements of intra-actions within an assemblage (Lucero & Garoian, 2016; Sellar, 2012). In Diya's case, although she in no way wanted to be part of nonconsensual sexual behavior and in no way perpetrated any crime, her presence allowed for the reterritorialization of violence through the event. Intra-actions involve sensation because, to borrow from Gershon (2019), we live sensual lives. Even violence and the sensations attached to it are a part of the sensuous nature of experiences. Diya's discussion of nonconsensual sexual relationships provides but one example of how the assemblage of violence normalizes toxic masculinity while demanding physical and emotional subservience from women. Think of the #MeToo movement, where masculine norms were interrupted on a global scale. Prior to #MeToo, while there were certainly movements to disrupt norms of "letting it happen", as Diya explained, there were few internationally recognized movements to interrupt rape culture.

Part of the difficulty in disrupting toxic masculinity is that even the most explicit forms of violence are normalized, as people learn such messages from everyday ordinary affects (Stewart, 2007). Yet the lesson is learned twice, first from feeling affects and second from watching the local and less-local responses to those affects. In other words, affects can be points of instruction. Returning to the example of Vanitha Mathil, the women's protest in India, there were several messages being sent and received that taught those on both sides about sociocultural roles. As women watched police officers dressed in their standard khaki uniforms, standing in rows with lathi sticks—batons that are vestiges of British colonial rule (Haines & Sarkar, 2020)—or rifles, fear was both about the context and about past, present, and future affects knotted within the assemblage. When scenes of citizens being brutally attacked with these weapons surface in the media in ways that are often largely unabated, fear becomes the lesson proliferated within the assemblage. Such fear is often reasonably carried into moments when police clash with citizens, because the memory of watching violence, of feeling that fear, becomes ingrained in a person's way of being and knowing.

Moments of explicit violence can thus be understood as a formal curriculum, one that a perpetrator intends to teach a victim. Within these moments of intentional aggression, affects are entangled with and across contexts. Like the baton, it is not just the physical reminder of fear but the fear itself that teaches us. To be clear, this discussion is not an attempt to flatten violence—there are, of course, moments of violence that are unintended but nonetheless hurtful and harmful to those on the receiving end.

What I focus on here are aggressions that are carried out with intention. In Kareena's narrative in the previous chapter, her husband and his family's violence intentionally taught Kareena about her sociocultural role within systems of gender and class.

This intentional teaching of a "woman's place" is echoed in many schools in India. Much like schools in Western nations, gender inequities are present in schools across the country (Majumdar & Mooij, 2011). The emotional violence against girls in Indian schools is explicit, and often intentional. Girls are often given different assignments than boys, expected to do labor like cleaning, and are far less likely to attend even "low fee" private schools (Majumdar & Mooij, 2011). These reflect an effort to reinforce gender inequities within socioculturally accepted roles for girls and women. In these cases, a formal curriculum of deference is intentionally enacted between adults and girls in school.

The null curriculum, or that which is explicitly not taught, is also present outside of schools. When Diya explains what she was taught through her entangled relationship with sociocultural norms and values, lessons of equity and access that might counter a curriculum of violence were missing. In the introduction, I offered an overview of Gershon's (2018) work on intention, attention, and expression. Both formal and null curricula are based in intention (whose narratives do those who write the curriculum want expressed?), attention (to whose narratives do we attend?), and expression (how are the stories expressed in a way that will maximize their impact?).

The null curriculum is often a question of sociocultural attention. While the intention might be to teach roles through aggression, there is also the question of which voices receive attention on local and less-local scales. Because of a lack of attention to voices decrying intimate partner violence, the silencing of women's rights groups becomes the null curriculum. As but one example, consider the outcry against normalized violence against African Americans in the United States as heard through the #BlackLivesMatter movement. This movement is incredibly significant because it stands against years of oppression through enslavement, Jim Crow, and white supremacy. One could understand #BlackLivesMatter as a formal curriculum delivered by activists, scholars, Black communities, and their accomplices to intentionally impress the value of Black lives against a country that has historically devalued and dehumanized Black people and groups.

Yet often absent from #BlackLivesMatter dialogues are the rampant murders of transwomen of color across the United States. This is true despite the fact that the #BlackLivesMatter movement was founded by queer Black women (Cohen & Jackson, 2016). Patrisse Cullors, one of the cofounders of Black Lives Matter, has similarly argued that while the movement has "worked to support Black trans women…our movement, collectively, has not been as present for [Black transwomen's] murders" (Tourjée, 2019, para. 7).

What do these silences indicate about the null curriculum of #BlackLivesMatter? Thinking critically about such absences is important because unpacking a movement so central to racial equity across the United States can help strengthen the movement's inclusivity, and can further disrupt white, cis-hetero norms and values. Even more significant is an examination of the echo chamber of null curricula between schools and communities. If, for example, Black trans voices are absent both from schools and from the movements that advocate for their rights within questions of race, what does a Black trans person learn about their own value from this compounded null curriculum? What does a transphobic white supremacist learn about the value of these transwomen?

Within the echo chamber of oppression, women and girls in India receive messages about their value through the null curricula in schools and through their communities. In schools, the null curriculum is predicated on an absence of attention to women's voices and choices, a message that is in close relationship to those found in communities (Allender, 2016). Women like Diya are therefore just as informed by what they were not explicitly taught as by what they were; they are taught by the absent voices of the null curriculum. Similarly, as these women are always already in intra-action with their local and less-local communities, their ways of being, knowing, and doing are informed by the enacted and hidden curriculum within those intra-actions.

Conclusion

A curriculum of violence normalizes aggressions until they become business as usual. This is true across contexts, from schools to communities. This particular curriculum is insidious, in that it is diffused across all forms of curriculum—formal, null, enacted, and hidden. Much like racism or misogyny are described as being "in the water", like lead or other dangerous contaminants, the ability to ignore this curriculum is a matter of privilege. This is like any other form of privilege that allows people to not have to attend oppression, while at the same time providing a kind of amnesty from being seen as abusing privilege when the person remains ignorant of the oppressions that impact others. Thinking with this metaphor further, the curriculum of violence is not a question of how much lead is in the water, but rather how much of the water one has to drink and whether or not you have the privilege of a filter. The lead—in this chapter, racism, sexism, and homo- and transphobias—is always in the water. Drinking water is necessary to live. But the ability to purchase or be given a filter for protection is a question of privilege. How one uses the filter, for example as an ally or accomplice for those who do not have their own filters, is a matter of agency. How much one can create their own filters is a question of agentic contingencies.

Assemblages can therefore be understood as pedagogical in nature. Their entanglements are pedagogical as bodies intra-act within violent norms and values. Perhaps one of the strongest tools of violent assemblages are capitals of shame, which function to keep the status quo in and across contexts. The movements of aggression within the curriculum of violence happen with friction and force. While this chapter has predominately focused on the movements caused by oppression, it is important to remember that both sides can be worn by such relationships. The #BlackLivesMatter movement brought national attention to the epidemic of maiming and killing Black people. While historical friction and force has narrowly positioned African Americans, Black Lives Matter still managed to raise national and international awareness of this epidemic. When friction and force are shifted, and the margins wear away at the malignancy of violent norms, social change toward equity is possible.

As COVID-19 continues to rage across the planet, secondary waves of illness are punctuated by joy and despair, kindness and violence. Although schools no longer involve face-to-face interactions, the curriculum of violence continues. Such lessons are not only taught in the schoolroom, they exist as a reciprocal relationship between the multiple places in which learning occurs. Interrupting the curriculum of violence, like disrupting any form of curriculum, requires an ethical fidelity to the attention, intention, and expression of antiviolent practices. For the students at New Horizons who did not find support at home, I can only imagine how difficult this time has been. For the women who lost their homes when shelters like Aasha closed due to health concerns, or for those now locked away with a steady onslaught of aggression, I am afraid to imagine their pain.

The significance of a curriculum of violence is not just that it exists, can be named, and can be traced across contexts. Scholars like Cooper (1892) and Woodson (1933), among others, have already articulated how violence is normalized, engendered, and maintained through educational processes. Building on these conversations, framing aggression within a curriculum means that it can also potentially be interrupted. For just as a curriculum of violence normalizes racism, sexism, homo- and transphobia, xenophobia, and the like, there are also antiracist (e.g., Giroux, 1992; Love, 2019; Minthorn et al., 2019), anti-queerphobic (e.g., Airton, 2018; Pinar, 2003), and feminist curricula (e.g., Pinar & Miller, 1982; Sabzalian, 2018) that, when used across contexts from schools to communities, are essential in disrupting the way that violence is a lesson too often presented as business as usual.

References

Abend, A. C. (2001). *Planning and designing for students with disabilities*. Washington, DC: Office of Educational Research and Improvement.

Agar, M. H. (1996). *The professional stranger: An informal introduction to ethnography* (2nd ed.). San Diego, CA: Academic Press.

Airton, L. (2018). *gender: Your Guide: A gender-friendly primer on what to know, what to say, and what to do in the new gender culture*. New York, NY: Simon and Schuster.

Allender, T. (2016). *Learning femininity in colonial India, 1820–1932*. Manchester, UK: Manchester University Press.

Anyon, J. (2000). Social class and hidden curriculum of work. In S. J. Ball (Ed.), *Sociology of education, vol. 2* (pp. 1250–1275). New York, NY: RoutledgeFalmer.

Apple, M. W. (1993). The politics of official knowledge: Does a national curriculum make sense? *Teachers College Record*, *95*(2), 222–241.

Bachman, J. S. (Ed.). (2019). *Cultural genocide: Law, politics, and global manifestations*. New York, NY: Routledge.

Barad, K. (2007). *Meeting the universe halfway: Quantum physics and the entanglement of matter and meaning*. Durham, NC: Duke University Press.

Barnes, M. & Gonzalez, J. (2015). *Hacking education: 10 quick fixes for every school*. Cleveland, OH: Times 10 Publications.

Bethune, M. M. (1938). Clarifying our vision with the facts. *Journal of Negro History*, *7*(2), 10–15.

Bobbitt, F. (1909). Practical eugenics. *The Pedagogical Seminary 16(3)*, 385–394.

Bobbitt, F. (1918). *The curriculum*. Boston, MA: Houghton Mifflin Company.

Bogart, N. (2020). Advocates scramble to help domestic abuse victims as calls skyrocket during COVID-19. *CTV News*. Retrieved from: https://www.ctvnews.ca/health/coronavirus/advocates-scramble-to-help-domestic-abuse-victims-as-calls-skyrocket-during-covid-19-1.4923109

Bourdieu, P. & Passeron, J.C. (1970). *Reproduction in education, society, and culture*. Thousand Oaks, CA: Sage Publications.

Burawoy, M. (2019). *Symbolic violence: Conversations with Bourdieu*. Durham, NC: Duke University Press.

Campbell, A.M. (2020). An increasing risk of family violence during the COVID-19 pandemic: Strengthening community collaborations to save lives, *Forensic Science International Reports*, *2 (89)*, doi:10.1016/j.fsir.2020.100089.

Chakraborty, R. (2020). Responding to the domestic violence crisis of COVID-19. *Harvard Center for Health and Human Rights*. Retrieved from: https://fxb.harvard.edu/2020/04/20/responding-to-the-domestic-violence-crisis-of-covid-19/

Cohen, C. J., & Jackson, S. J. (2016). Ask a feminist: A conversation with Cathy J. Cohen on Black Lives Matter, feminism, and contemporary activism. *Signs: Journal of Women in Culture and Society*, *41*(4), 775–792.

Cooper, A. J. (1892). *A voice from the south (by a black woman from the south)*. Xenia: OH: Aldine Printing House.

Crichlow, W. (2002). Western colonization as disease: Native adoption and cultural genocide. *Critical Social Work*, *2*(2), 104–127.

Dashow, J. (2018). New FBI statistics show alarming increase in number of reported hate crimes, *Human Rights Organization*. Retrieved from: https://www.hrc.org/blog/new-fbi-statistics-show-alarming-increase-in-number-of-reported-hate-crimes

Department of Justice. (2018). FBI releases 2018 hate crime statistics. The United States Department of Justice. Retrieved from: https://www.justice.gov/hatecrimes/hate-crime-statistics.

Dewey, J. (1938). *Experience and education*. Philadelphia, PA: Kappa Delta Pi.

DuBois, W.E.B. (1903). *The souls of black folk*. Chicago, IL: McClurgand Co.

Eisner, E. (1985). *The educational imagination: On the design and evaluation of school programs* (2nd ed.). New York, NY: Macmillan.

Ellsworth, E. (2005). *Places of learning: Media, architecture, and pedagogy*. New York, NY: RoutledgeFalmer.

Erickson, F. (1982). Taught cognitive learning in its immediate environments: A neglected topic in the anthropology of education 1. *Anthropology & Education Quarterly*, *13*(2), 149–180.

Fanon, F. (1961). *The wretched of the earth*. New York, NY: Grove Press.

Farley, L. (2018). *Childhood beyond pathology: A psychoanalytic study of development and diagnosis*. New York, NY: SUNY Press.

Febve, R. (2020). Interview with France's minister for gender equality. *France24*. Retrieved from: https://www.france24.com/en/20200421-french-gender-equality-minister-on-helping-victims-of-domestic-violence-during-lockdown.

Feld, S., & Brenneis, D. (2004). Doing anthropology in sound. *American Ethnologist*, 31(4), 461–474.

Foucault, M. (1978). *The history of sexuality*. New York, NY: Pantheon Books.

Freire, P. (1970). *Pedagogy of the oppressed*. New York, NY: Continuum Books.

Gay, G. (2002). Preparing for culturally responsive teaching. *Journal of Teacher Education*, *53*(2), 106–116.

Gershon, W. S. (2013). Sonic cartography: Mapping space, place, race, and identity in an urban middle school. *Taboo: The Journal of Culture and Education*, *13*(1), 4.

Gershon, W.S. (2017). *Sound curriculum: Sonic studies in educational theory, method, & practice*. New York, NY: Routledge.

Gershon, W. S. (2018). Reverberations and reverb: Sound possibilities for narrative, creativity, and critique. *Qualitative Inquiry*, 1–11.

Gershon, W.S. (2019). *Sensuous curriculum: Politics and the senses of education*. Charlotte, NC: Information Age Publishing Inc.

Gilbert, J. (2014). *Sexuality in school: The limits of education*. Minneapolis, MN: University of Minnesota Press.

Giroux, H. A. (1992). Post-colonial ruptures and democratic possibilities: Multiculturalism as anti-racist pedagogy. *Cultural Critique*, (21), 5–39.

Giroux, H. A. (2000). Public pedagogy as cultural politics: Stuart Hall and the crisis of culture. *Cultural Studies*, *14*(2), 341–360.

Giroux, H. (2018). *Pedagogy and the politics of hope: Theory, culture, and schooling: A critical reader*. New York, NY: Routledge.

Giroux, H. & A. Penna (1983) *The hidden curriculum and moral education*. Berkeley, CA: McCutchan.

Gottfried, M. A., Stiefel, L., Schwartz, A. E., & Hopkins, B. (2019). Showing up: Disparities in chronic absenteeism between students with and without disabilities in traditional public schools. *Teachers College Record*, *121*(8), 1–34.

Green, M. (1978). *Landscapes of learning*. New York, NY: Teachers College Press.

Haines, M. B., & Sarkar, S. (2020). Sticks, stones, and the secular bones of Indian democracy. *Engaging Science, Technology, and Society*, *6*, 133–141.

Halberstam, J. (1993). Imagined violence/queer violence: Representation, rage, and resistance. *Social Text*, (37), 187–201.

Haritaworn, J., Kuntsman, A., & Posocco, S. (Eds.). (2014). *Queer necropolitics*. New York, NY: Routledge.

Helfenbein Jr, R. J. (2004). A radical geography: Curriculum theory, performance, and landscape. *Journal of Curriculum Theorizing*, *20*(3), 67–75.

Hooks, B. (2003). *Teaching community: A pedagogy of hope*. New York, NY: Routledge.

Kovach, M. (2013). Treaties, truths and transgressive pedagogies: Re-imagining indigenous presence in the classroom. *Socialist Studies/Études Socialistes*, *9*(1), 109–127.

Kingdon, G. G. (2020). The private schooling phenomenon in India: A review. *The Journal of Development Studies*, 1–23.

Kitchin, R., Perkins, C., & Dodge, M. (2009). *Rethinking maps: New frontiers in cartographic theory*. New York, NY: Routledge.

Kozol, J. (2005). *The shame of the nation: The restoration of apartheid schooling in America*. New York, NY: Broadway Books.

Ladson-Billings, G. (1998). Just what is critical race theory and what's it doing in a nice field like education?. International journal of qualitative studies in education, 11(1), 7–24.

Levesque, K., Lauen, D., Teitelbaum, P., Alt, M., & Librera, S. (2000). Vocational Education in the United States: Toward the Year 2000. Statistical Analysis Report.

Love, B. L. (2019). *We want to do more than survive: Abolitionist teaching and the pursuit of educational freedom*. Boston, MA: Beacon Press.

Lucero, J., & Garoian, C. (2016). The performative antics of assemblage pedagogy. *Visual Inquiry: Learning & Teaching Art*, *5*(3), 451–455.

Lusted, D. (1986). Why pedagogy? *Screen*, *27(5)*, 2–14.

Lytvyn, A., Lytvyn, V., Rudenko, L., Pelekh, Y., DIdenko, O., Muszkieta, R., & Żukow, W. (2020). Informatization of technical vocational schools: Theoretical foundations and practical approaches. *Education and Information Technologies*, *25*(1), 583–609.

Majumdar, M., & Mooij, J. E. (2011). *Education and inequality in India: A classroom view* (Vol. 46). New York, NY: Routledge.

Massumi, B. (2002). *Parables for the virtual: Movement, affect, sensation*. Durham, NC: Duke University Press.

Mbembe, A. (2019). *Necropolitics*. Durham, NC: Duke University Press.

McCready, L. T. (2004). Some challenges facing queer youth programs in urban high schools: Racial segregation and de-normalizing whiteness. *Journal of Gay & Lesbian Issues in Education*, *1*(3), 37–51.

Mehan, H. (1979). *Learning lessons*. Cambridge, MA: Harvard University Press.

Metz, M. H. (1978). *Classrooms and corridors: The crisis of authority in desegregated secondary schools*. Los Angeles, CA: University of California Press.

Milner, H. R. (2011). Culturally relevant pedagogy in a diverse urban classroom. *The Urban Review*, *43*(1), 66–89.

Milner, H.R. (2017). Confronting inequity/reimagining the null curriculum. *Educational Leadership 75(3)*, 88–89.

Ministerio de Igualdad, Gobierno de España. (2020). *Delegación del gobierno contra la violencia de género: Por una sociedad libre de violencia de género*. Retrieved from: https://violenciagenero.igualdad.gob.es/violenciaEnCifras/home.htm

Minthorn, R. Z. T. H. A., Belone, L., Martinez, G., & Sims, C. (2019). Indigenous research perspectives in the State of New Mexico: Implications for working with schools and communities. *Journal of American Indian Education*, *58*(1–2), 108–123.

Morris, M.W. (2016). *Pushout: The criminalization of black girls in schools*. New York, NY: The New Press.

Myers, C.Y. (2019). *Children and materialities: The force of the more-than-human in children's classroom lives*. New York, NY: Springer.

National Center for Education Statistics (2019). The Inclusion of students with disabilities. Retrieved from: https://nces.ed.gov/fastfacts/display.asp?id=59

Nespor, J. (1997). *Tangled up in school: Politics, space, bodies and signs in the educational process*. Mahwah, NJ: Lawrence Erlbaum Associates Inc.

Oakes, J. (1985). *Keeping track: How schools structure inequality*. New Haven, CT: Yale University.

Oliveros, P. (2005). *Deep listening: A composer's sound practice*. Lincoln, NE: Deep Listening Publications.

Page, R. (1991). *Lower-track classrooms: A curricular and cultural perspective*. New York, NY: Teachers College Press.

Paris, D. (2012). Culturally sustaining pedagogy: A needed change in stance, terminology, and practice. *Educational researcher*, *41*(3), 93–97.

Patel, L. (2015). *Decolonizing educational research: From ownership to answerability*. New York, NY: Routledge.

Pinar, W. F. (2003). Queer theory in education. *Journal of Homosexuality*, *45*(2–4), 357–360.

Pinar, W. F., & Miller, J. L. (1982). Feminist curriculum theory: Notes on the American field 1982. *The Journal of Educational Thought (JET)/Revue de la Pensée Educative*, *16*(3), 217–224.

Pinch, T., & Bijsterveld, K. (Eds.). (2012). *The Oxford handbook of sound studies*. New York, NY: Oxford University Press.

Powell, K. (2010). Making sense of place: Mapping as a multisensory research method. *Qualitative Inquiry*, *16*(7), 539–555.

Ratnam, D. (2020). Domestic violence during COVID-19 lockdown emerges as serious concern, *Hindustan Times*. Retrieved from: https://www.hindustantimes.com/india-news/domestic-violence-during-covid-19-lockdown-emerges-as-serious-concern/story-mMRq3NnnFvOehgLOOPpe8J.html

Refuge. (2020). *Refuge, the UK's largest domestic abuse charity says calls to its helpline have risen by 25% since lockdown measures began.* Retrieved from: https://www.refuge.org.uk/25-increase-in-calls-to-national-domestic-abuse-helpline-since-lockdown-measures-began/

Roy, D., Baker, W., & Hamilton, A. (2019). *Teaching the arts*. Cambridge, MA: Cambridge University Press.

Sabzalian, L. (2018). Curricular standpoints and native feminist theories: Why native feminist theories should matter to curriculum studies. *Curriculum Inquiry*, *48*(3), 359–382.

Sabzalian, L. (2019). *Indigenous children's survivance in public schools*. New York, NY: Routledge.

Sandlin, J. A., Schultz, B. D., & Burdick, J. (Eds.). (2010). *Handbook of public pedagogy: Education and learning beyond schooling*. New York, NY: Routledge.

Schafer, R. M. (1969). *The new soundscape*. Don Mills, Canada: BMI Canada Limited.

Schwab, J. J. (1969). The practical: A language for curriculum. *The School Review*, *78*(1), 1–24.

Sedgwick, E. K. (1993). *Touching feeling: Affect, pedagogy, performance*. Durham, NC: Duke University Press.

Sedgwick, E. K. (2003). *Touching feeling: Affect, pedagogy, performance*. Durham, NC: Duke University Press.

Sellar, S. (2012). 'It's all about relationships': Hesitation, friendship and pedagogical assemblage. *Discourse: Studies in the Cultural Politics of Education*, *33*(1), 61–74.

Simpson, A. (2014). *Mohawk interruptus: Political life across the borders of settler states*. Durham, NC: Duke University Press.

Smith, L. T., Tuck, E., & Yang, K. W. (Eds.). (2018). *Indigenous and decolonizing studies in education: Mapping the long view*. New York, NY: Routledge.

Spencer, H. (1859). What knowledge is of most worth? *The Westminster Review*, *16(1)*, 522–532.

Stewart, K. (2007). *Ordinary affects*. Durham, NC: Duke University Press.

Tatum, B. D. (2003). *"Why are all the Black kids sitting together in the cafeteria?": And other conversations about race*. New York, NY: Basic Books.

Thomas, D. (2019). Black lives matter as resistance to systemic anti-Black violence. *Journal of Critical Thought and Praxis*, *8*(1), 4.

Tourjée, D. (2019). The killing of transwomen is inseparable from Black Lives Matter, *Vice: Identity.,* Retrieved from: https://www.vice.com/en_us/article/neaa5g/transwomen-black-lives-matter

Tsing, A.L. (2005). *Friction: An ethnography on global connection*. Princeton, NJ: Princeton University Press.

Valente, J.M. (2011). *d/Deaf and d/Dumb: A portrait of a deaf kind as a young superhero*. New York, NY: Peter Lang Publishing.

Varenne, H., & McDermott, R. (1998). *Successful failure: The school America builds*. Boulder, CO: Westview.

Visweswaran, K. (2010). *Un/common cultures: Racism and the rearticulation of cultural difference*. Durham, NC: Duke University Press.

Wang, H. (2014). *Nonviolence and education: Cross-cultural pathways*. New York, NY: Routledge.

Watkins, W. H. (2001). *The white architects of black education: Ideology and power in America 1865–1954*. New York, NY: Teachers College Press.

Wiggins, J. & McTighe, G. (2005). *Understanding by design*. Alexandria, VA: Association for Supervision and Curriculum Development.

Winfield, A. G. (2007). *Eugenics and education in America: Institutionalized racism and the implication of history, ideology, and memory*. New York, NY: Peter Lang.

Wozolek, B., Varndell, R., & Speer, T. (2015). Are we not fatigued?: Queer battle fatigue at the intersection of heteronormative culture. *International Journal of Curriculum and Social Justice, 1(1),* 186–214.

Wozolek, B., Wootton, L., & Demlow, A. (2017). The school-to-coffin pipeline: Queer youth, suicide and resilience of spirit. *Cultural Studies<>Critical Methodologies, 17(5*), 392–398.

Yoon, I. H. (2019). Haunted trauma narratives of inclusion, race, and disability in a school community. *Educational Studies*, *55*(4), 420–435.

CONCLUSION

Reflecting On The Assemblage

When girls' night is over and I've cleaned up sticky bowls of ice cream and turned off the TV, I often quietly go into my children's room to say goodnight one last time. With any luck, both kids will be fast asleep and will stay that way for the next few hours. Like most parents, the sight of my sleeping babies is met with joy and exhaustion, the desire to kiss them gently while fearing that any movement will wake them up. I tuck them back into their blankets, making sure they are warm and wishing I could preserve them as they are right now. I know, however, that this moment will fade into a soft memory, like so many others.

One night, as I walked softly into the room, begging the floorboards not to creak, my daughter slowly lifted her head and asked, "Are you done?"

"With what, baby?" I asked.

"With work. You always go back to work. Are you done?" she asked in a sleepy voice.

"Not quite, baby", I whispered. "There's much to be done. Go to sleep so you don't wake your brother". I kissed her on the head and, as she had anticipated, returned to the office to write.

Sitting in the office and thinking about all the implications of violence, I couldn't help but wonder how my own children will be entangled in it. My daughter, with her gorgeous brown skin, hair, and eyes, will likely experience similar forms of oppression to those I have lived. My son, with his beautiful blonde hair, light skin, and blue eyes will likely experience far more privilege than I could have imagined. As I considered the work to be done with this text, I found myself thinking about ways to protect them both, because, in the end, both will be in some way entangled within the assemblage of violence. They will learn from violence in ways that I cannot control. They will be formed and informed by it.

The more I thought about the possible trajectories set before them, the more I could feel those chains on my skin and between my fingers, cold metal lingering on my body. I am entangled with my own past, present, and future, but I am also inextricably knotted with theirs. When my children feel pain, part of me feels it too, and I wonder: As they become increasingly aware of the world, how much of my pain do they feel? Perhaps more importantly, is it *my* pain or is it *ours*?

Before I knew it, my mind had wandered and I was sitting with a montage of memories—the father of a boy I liked in high school encouraging me to have sex with his son so he could enjoy someone "exotic" before settling down with a "nice girl", a boy after the 9/11 attacks asking me if my dad was a "towel head", children in kindergarten explaining that I was too "dirty" to use the peach crayon, a customer at the grocery store telling me to "take [my] spic ass back to Mexico", a scholar telling me that my skin is "too light" to understand oppression, a boyfriend slamming me against his parents' van when I told him the relationship was over and holding me there while he kissed me, and countless memories of a relationship that was punctuated by sexual assault.

I realized in recounting these memories that I had paused. The cursor was still blinking on the screen, calling to me like a mother impatiently tapping her foot, waiting for her child to listen. As scholars, we tend to engage in stories that are meaningful to us. My work at Aasha and New Horizons resonates with my own experiences with violence. But more than this, they are also narratives with which I hope my own children never have to intra-act. My mind wandered some more, and I found myself thinking about the eight years that my daughter had been on this planet. I found myself recalling the time that she cried at me not to "speak Spanish to [her] ever again because it makes [her] 'weird'", or the time that she told me to only pack "normal" lunch items and not Indian snacks. I wonder how many experiences she has already had with violence that, like me, she holds in her heart, too afraid to burden her parents with these truths.

Finally, my mind arrived back at New Horizons. I remembered one sunny afternoon when I sat with some of the girls from the Blue Dragons outside the chicken coop. After a moment, I had noticed that Zella, a thin African American seventh grader, was staring at me awkwardly from above her thick, black-framed glasses. When I asked if I could help her with something, she said, "How do you keep answers in your head? I mean, when the teacher calls on you. This Spanish teacher, Señor Rodriguez, makes my mind go blank every time he calls on me! Then I get an attitude because it makes me feel bad. And you know what he does? He tells me to leave. Every time. The worst part is that *I knew the answer!* But I doubt myself. Maybe I just need a way to make other people think I'm smart. Is there a way to do that? I don't know how to do that either…" Her voice trailed off, becoming pensive and sullen. The question hit me so unexpectedly that I paused to consider the most impactful way to answer her.

After a minute, I realized that the other girls were also leaning forward as if to hear some sage words of wisdom. "I don't know", I answered. "I'll be honest. I have a doctorate, which means I studied for 12 years after high school and sometimes I still feel surprised when someone tells me I'm smart".

Zella laughed, "Twelve years?!? You were in school for as long as I've been alive!"

"Okay, Zella! Stop doing the math…. Anyway", I continued, "Here's the thing, I guess you just have to learn to know it for yourself and not care what anyone else thinks. It's like one of my friends once told me, Beyoncé doesn't have to tell other people that she's a queen, right? So, if you know it, they will know it".

"That's it?" asked Zella, clearly looking for a one-liner that I couldn't produce.

"Sorry. That's all I've got. But I can tell you this—every single one of you is incredibly smart and talented. I know because I've been hanging out with you for almost two years now. Trust me, I know these things". In the way that many adolescent conversations go, the topic quickly changed without a trace of the previous dialogue. But when Zella went to leave that day, she just looked at me and said, "You sure about us?" I nodded as she smiled and ran off.

Later that same afternoon, I engaged in one of my writing rituals—the one where I stare at a blank page and give in momentarily to the deep-seated fear that my writing won't be smart or interesting enough in the court of academic approval. Imposter syndrome (Clarce & Imes, 1978), or self-perceived intellectual doubt, is not unusual to academics. Although this phenomenon impacts people across contexts, marginalized people are far more likely to engage with this experience than their peers who belong to dominant groups (Edwards, 2019; González, 2020; Jones, 2017; Williams, 2016). As I sat there, looking intensely at that page, my mind wandered back to the conversation with Zella earlier that day. What does it mean that imposter syndrome is a normal part of my academic existence? Perhaps more significantly, what does it mean that a 12-year-old queer Black girl was wrestling with similar doubt? To better answer these questions, it is necessary to further theorize and contextualize imposter syndrome.

Rather than an internally constructed anxiety, I theorize imposter syndrome as an expression of an assemblage of violence. Assemblages of violence are the many iterations of past, present, and future aggressions that are entangled across sociocultural and political contexts. In other words, any one articulation of violence is necessarily dependent on other violent events. As I have argued throughout this book, this is not an attempt to flatten the dynamic and deeply contextualized nature of violent acts. It is, however, an argument for paying attention to the knotted nature of aggression.

Clarce and Imes (1978) first coined the term "imposter phenomena" after working with 150 women over a 5-year period. Every woman represented in their study had earned PhDs and produced numerous publications for top-tier journals. Many had received nationally recognized awards for their work, and all had been acknowledged with various forms of praise from colleagues. By academic standards, these women might be considered "highly successful", a term used by Clarce and Imes in describing their participants. Yet all of the women admitted that they did not experience internal feelings of success. This sensation of doubt lead to anxiety, stress, and depression in several participants. Clarce and Imes noted anecdotally in their practices as psychologists that men were far less likely to experience imposter phenomena.

Clarce and Imes' work is significant because it was the first to document imposter syndrome along with its sociopolitical and cultural underpinnings, which too often leave people feeling inadequate. While this study focused on white women from middle- to upper-middle-class backgrounds, it provided the language for describing the outcome of broader marginalizations within the academy. Since their study, several scholars have noted that imposter syndrome is often intensified for those whose ways of being reside at the intersections of marginalized positionalities (e.g., Edwards, 2019; González, 2020; Jones, 2017; Williams, 2016). This is because the marginalization of identities exists like an echo chamber, with violence crashing continuously onto multiple ways of being, knowing, and doing. The more intersecting identities a person holds that are marginalized, the greater the possibility that violent reverberations will negatively impact that person.

To be clear, I am not arguing that people should engage in some kind of oppression Olympics. Nor do I condone the kinds of identity politics that are akin to colorism or respectability politics. Neither of these are particularly useful for interrupting oppression, and too often end up reifying violent norms in and between groups. What I am arguing for is an attention to the multiple iterations of violence that tend to be in broad and pointed circulation around marginalized people and groups. I also argue that we need to pay attention to the agency that is coconstituted within echo chambers of oppression, as well as how one might react to violence. In other words, what might be read as similar forms of aggression might impact people in radically different ways, in part because of how they are affected and in part because of the agency that was constituted through violent intra-actions.

Imposter syndrome is therefore complex not only because of its relationship with past, present, and future aggressions, but also because of the multiplicity with which it proliferates within intention (what is the intention of violence?), attention (to what does the aggressor attend when they exhibit violence?), and expression (how is violence expressed?; Gershon, 2018). The person who experiences imposter syndrome is responding to external messages that are intended to maintain marginalization. In Zella's case, when she was called on or during tests, her answers disappeared as a stress response to the context that was nested in broader messages of inadequacy. As Barad (2007) argues, within a phenomenon such as an act of aggression, bodies are in constant intra-action. This means that aspects like meaning making and agency are coconstituted through the event, rather than separate from it. Everyday acts of aggression can therefore lead to imposter syndrome, but it is not just that a person experiences violence but rather that she is in intra-action with violence, and therefore in relationship to questions of intention, attention, and expression. Imposter syndrome is not just an act of self-doubt caused by broader acts of violence, it can also be understood as emotional self-harm.

The feedback loop that fuels imposter syndrome is therefore both about external messages and about how those messages are received. In other words, the same questions of intention, attention, and expression are also resonant with the self-harms engendered through violent intra-actions. Survivors often undergo a process of evaluating their aggressor's intentions, and sometimes their own. This evaluation is twofold: On one hand, she might second guess her own intentions within a violent event; Zella, for example, might ask herself if she wanted to leave and therefore found herself avoiding the teacher's question. On the other hand, there are sometimes questions—albeit subconscious ones—about the intention of the self-harm. Within imposter syndrome, a survivor might ask why she allows herself to feel crushing self-doubt. Similarly, she might attend to different moments of the event, and as a result, attend to her self-image differently. Finally, imposter syndrome, like many forms of self-harm, has multiple forms of expression.

Yet, as complex as imposter syndrome might be, it is but one expression of an assemblage of violence. Such expressions often bubble up or erupt from the assemblage, materializing as an event that is singular or part of a broader pattern—like the inability to answer a question in class or an insecurity about writing—and then receding back into the assemblage. The eruption itself deterritorializes the assemblage as violence emerges through an event. However, as that act of aggression folds back into the assemblage, it reterritorializes the assemblage; that seemingly singular event becomes a part of the past, present, and future iterations of violence.

An assemblage can be understood as a knot of entanglements. The entanglements of an assemblage are one way to think about the infinite number of relations and relationships in and outside of any ecology that cannot be pulled apart. When violence erupts from an assemblage, the relationships that are central to it are not destroyed through deterritorialization, but rather are shifted by the intra-actions of violence. When a queer woman of color experiences gaslighting based on her gender, race, or sexual orientation, she is intra-acting with violence. In Zella's case, doubt was cast on her ability to, in her words, "be smart". This was accomplished not in a single moment but through a multiplicity of imbricated events and contexts that made her feel insecure about her intellectual abilities. Over time, this culminated in imposter syndrome. As Zella experienced violence that morphed from external messages to internal harm, the singular events changed and reinforced the broader assemblage through its reterritorialization.

The entanglements within an ever-changing assemblage are also significant in that sometimes bodies become entangled on the quantum level. Quantum entanglements are when two bodies—human or nonhuman—become so deeply linked that they can impact each other across distance. As traditionally framed, when one body of an entanglement is measured, the other changes in reaction to that measurement. In quantum mechanics, for example, if one photon has an up spin upon measurement, the other will have a down spin. In the case of New Horizons students at the talent show, hope entangled with hope. From both sociological and even recent scientific perspectives, the idea of quantum entanglement should be expanded beyond only opposite reactions to account for the interplay of like affects, like hope.

Quantum entanglements can occur between people, as they did with hope at New Horizons, but they can also occur between human and nonhuman bodies. Similar to Prisha's narrative in Chapter 1, in which her experiences entangled her ways of being, knowing, and doing with the "body" of culture, here I argue that when a person experiences imposter syndrome, she is not just responding to marginalizing events and their values, she is entangled with them at the quantum level. This is what makes imposter syndrome so insidious. As queer girls of color like Zella gain confidence, broader sociocultural agents might work to further devalue and dehumanize her, engendering and maintaining dominant norms and values and causing new iterations of imposter syndrome. For queer people like Zella, continuous cycles of aggression often lead to queer battle fatigue (Wozolek et al., 2015) and, sometimes, to the school-to-coffin pipeline (Wozolek et al., 2017).

Another significant factor in violent intra-actions is the way some people use capitals of shame as a mechanism of control. Unlike other dialogues on cultural capital (e.g., Bourdieu & Passeron, 1977), shame is not something that one wishes to have but rather something that one wishes to use to gain social mobility. The women at Aasha and the students at New Horizons all describe various degrees of shame that impacted their ontoepistemologies, ranging across shame based on class, gender, gender identity and expression, sexual orientation, and race. As Zella's teachers and administrators became complicit in shame (e.g., "Don't shame the principal"), she became further entrenched in doubt, which carried into her feelings of incompetence.

Within what Barad (2007) calls intra-actions, whether they produce quantum entanglements or not, agency is always coconstituted within the event rather than separate from it. Because power is always already at play within intra-actions along with agency, one's agentic contingencies are also engendered and maintained through the phenomenon. Agentic contingencies are the many factors that limit one's ability to enact agency, the things on which one's mobility is contingent. For participants across the studies presented in this book, their agency was often limited by the same sociopolitical norms and values within the cis-heteropatriarchy that were fodder for shame, like race and gender. In Zella's case, or in any case of imposter syndrome for that matter, she became limited not only by succumbing to the aggressions she faced on a daily basis, but also through the internal messages that became the self-fulfilling prophecy of the negative cultural narrative she received in local and less-local contexts.

Finally, I note what is learned through an assemblage of violence. The curriculum of violence is a knot (Nespor, 1997) of local and less-local understandings that stems from the nested and layered nature of the formal, hidden, null, and enacted curricula. Although some educational theorists have discussed these forms of curriculum in and across contexts of schooling, I have argued that the curriculum of violence speaks and leaks (Helfenbein, 2010) from communities to the classroom and back again. In the example of imposter syndrome, the formal curriculum, what is meant to be taught (Apple, 1993), includes the aggressions that spur insecurity. The null curriculum, or what is explicitly not taught (Eisner, 1985), is often composed of messages that might interrupt self-fulfilling prophecies. These two forms of curricula both form and inform the hidden curriculum (Jackson, 1968), or the underlying norms and values that remain hidden from those entrenched in a culture. This is why imposter syndrome is so hard to name: it is rooted in a broader curriculum of violence that is so normalized through its formal iterations that it becomes hidden to those experiencing it. All of these forms of curriculum are part of the enacted curriculum (Page, 1991), or what we learn through the intra-actions of bodies in and across contexts.

The complex nature of violence makes its movements and trajectories hard to trace. I have suggested that pathology can be used as a method to trace these trajectories, and perhaps to predict their future movements as a tool for interrupting violence. Thinking with pathology can be helpful in unpacking the margins that are constructed around oppressions. However, pathology as a method should be taken on with the same kind of care and attention that other critical qualitative methodologies demand (e.g., Denzin & Giardina, 2009; Gershon, 2018; Rodriguez et al., 2016; Solóranzo & Yosso, 2002), in particular because of the historical sociopolitical violence surrounding research in the social and, in particular, natural sciences.

In the end, there are few ways to escape certain forms of sociopolitically and culturally normalized violence. While they have privilege and power, those who perpetuate it are still entangled by it. Those who are marginalized through it are ensnarled in often untenable yet inescapable ways. Yet, despite being in constant and continuous relation with violence through the assemblage, resistance, refusal, and resilience are all tools that survivors of aggression have at their disposal to varying degrees. Mahmood (2008) writes, "What courage is not possible when the courage of all people around us wells up…again and again until we are tired of watching it on our televisions and reading it in our newspapers—and yet people still find the strength to resist, and resolve and forage on ahead" (p. 10)? Using their agency despite contingencies, the participants in these studies used courage, resistance, refusal, and resilience as tools against oppression and to be their authentic selves.

Because violence is so pervasive, I can only assume that at least one person who reads this book has experienced some form of violence. The question is whether or not you had the privilege of ignoring it. Whether physical or emotional, any aggression is unacceptable. So, to those living with and within the assemblage of violence, this book is for you too. Your ability to use tools against oppression are recognized. Even if you've attempted to move away, physically or emotionally, from violence and are not as far in your progress of extricating yourself as you might like, your work is recognized. For, as much as violence erupts and changes an assemblage and shapes the way we learn and exist in the world, so too do hope, courage, and love.

References

Apple, M. W. (1993). The politics of official knowledge: Does a national curriculum make sense? *Teachers College Record, 95*(2), 222–241.

Barad, K. (2007). *Meeting the universe halfway: Quantum physics and the entanglement of matter and meaning*. Durham, NC: Duke University Press.

Bourdieu, P., & Passeron, J. C. (1977). *Reproduction in education, society and culture*. London, UK: Sage Publications.

Clarce, P.P., & Imes, S.A. (1978). The imposter phenomenon in high achieving women: Dynamics and therapeutic intervention, *Psychotherapy: Theory, Research & Practice, 15(3),* 241–247.

Denzin, N. K., & Giardina, M. D. (2009). *Qualitative inquiry and social justice: Toward a politics of hope.* Walnut Creek, CA: Left Coast Press.

Edwards, C. W. (2019). Overcoming imposter syndrome and stereotype threat: Reconceptualizing the definition of a scholar. *Taboo: The Journal of Culture and Education, 18*(1), 3.

Eisner, E. (1985). *The educational imagination: On the design and evaluation of school programs* (2nd ed.). New York, NY: Macmillan.

Gershon, W. S. (2018). Reverberations and reverb: Sound possibilities for narrative, creativity, and critique. *Qualitative Inquiry*, 1–11.

González, M. M. (2020). Queer Battle fatigue, or how I learned to stop worrying and love the imposter inside Me. *GLQ: A Journal of Lesbian and Gay Studies, 26*(2), 236–238.

Helfenbein, R. J. (2010). Thinking through scale: Critical geographies and curriculum spaces. In E. Malewski (Ed.), *Curriculum studies handbook: The next moment* (pp. 304–317). New York, NY: Routledge.

Jackson, P. W. (1968). *Life in classrooms.* New York, NY: Holt, Rinehart & Winston.

Jones, S. G. (2017). Death of the imposter syndrome: Scholarship through mentorship. In D. Y. Ford, M. T. Scott, R. B. Goings, T. T. Wingfield, & M. S. Henfield (Eds.), *RACE mentoring through social media: Black and Hispanic scholars share their journey in the academy* (pp. 163–170). Charlotte, NC: Information Age Press.

Mahmood, C. K. (2008). Anthropology from the bones: A memoir of fieldwork, survival, and commitment. *Anthropology and Humanism, 33(1/2),* 1–11.

Nespor, J. (1997). *Tangled up in school: Politics, space, bodies and signs in the educational process.* Mahwah, NJ: Lawrence Erlbaum Associates Inc.

Page, R. (1991). *Lower-track classrooms: A curricular and cultural perspective.* New York, NY: Teachers College Press.

Rodriguez, N. M., Martino, W. J., Ingrey, J. C., & Brockenbrough, E. (2016). *Critical concepts in queer studies and education.* New York, NY: Palgrave Macmillan.

Solórzano, D. G., & Yosso, T. J. (2002). Critical race methodology: Counter-storytelling as an analytical framework for education research. Qualitative inquiry, 8(1), 23–44.

Williams, B. C. (2016). Radical honesty: Truth-telling as pedagogy for working through shame in academic spaces. In F. Tuitt, C. Haynes, & S. Stewart (Eds.), *Race, equity, and the learning environment: The continued search for critical and inclusive pedagogies around the globe* (pp. 71–82). Sterling, VA: Stylus Publishing.

Wozolek, B., Varndell, R., & Speer, T. (2015). Are we not fatigued?: Queer battle fatigue at the intersection of heteronormative culture. *International Journal of Curriculum and Social Justice, 1(1),* 186–214.

Wozolek, B., Wootton, L., & Demlow, A. (2017). The school-to-coffin pipeline: Queer youth, suicide and resilience of spirit. *Cultural Studies<->Critical Methodologies, 17(5)*, 392–398.

INDEX

For Product Safety Concerns and Information please contact our EU representative GPSR@taylorandfrancis.com
Taylor & Francis Verlag GmbH, Kaufingerstraße 24, 80331 München, Germany

www.ingramcontent.com/pod-product-compliance
Ingram Content Group UK Ltd.
Pitfield, Milton Keynes, MK11 3LW, UK
UKHW021958180926
475701UK00003B/83

9780367409807